NETWORK NATURE

NETWORK NATURE

THE PLACE OF NATURE IN THE DIGITAL AGE

Richard Coyne

BLOOMSBURY ACADEMIC
LONDON • NEW YORK • OXFORD • NEW DELHI • SYDNEY

BLOOMSBURY ACADEMIC
Bloomsbury Publishing Plc
50 Bedford Square, London, WC1B 3DP, UK

BLOOMSBURY, BLOOMSBURY ACADEMIC and the Diana logo are trademarks of Bloomsbury Publishing Plc

First published in Great Britain 2018

A catalogue record for this book is available from the British Library.

Library of Congress Cataloging-in-Publication Data
Names: Coyne, Richard, author.
Title: Network nature: the place of nature in the digital age / Richard Coyne.
Description: London, UK; New York, NY, USA: Bloomsbury Academic, 2018. | Includes bibliographical references and index.
Identifiers: LCCN 2017042948 (print) | LCCN 2017044617 (ebook) | ISBN 9781350029514 (ePub) | ISBN 9781350029491 (ePDF) | ISBN 9781350029521 (hardback: alk. paper)
Subjects: LCSH: Philosophy of nature. | Human beings – Effect of environment on. | Digital media – Psychological aspects. | Virtual reality. | Semiotics.
Classification: LCC BD581 (ebook) | LCC BD581 .C687 2018 (print) | DDC 113 – dc23
LC record available at https://lccn.loc.gov/2017042948

ISBN: HB: 978-1-3500-2952-1
ePDF: 978-1-3500-2949-1
eBook: 978-1-3500-2951-4

Typeset by Deanta Global Publishing Services, Chennai, India
Printed and bound in Great Britain

To find out more about our authors and books visit www.bloomsbury.com and sign up for our newsletters.

To Chris

CONTENTS

PREFACE

Nature is on the side of the independent, the hopeful, the free, the good and the healthy. Who does not hanker after places and things that are unmediated, authentic and *natural*? But some digital-device users think that their technologies get in the way of direct access to nature. It is as if relentless connectivity, accompanied by work stress, boredom and poor health, burdens urban dwellers who must now look to *nature* to deliver the opposites of this technological affliction. It is easy to succumb to the view that nature is what is left in the crucible of human experience purged of troublesome technology and artifice.

But technology provides obvious benefits. Techno-science monitors, records and predicts the state of the natural world, from weather to life beneath the Arctic, and to great effect. Now you cannot make buildings of any scale without digitally enhanced surveys of sites, terrains, local bio-ecologies and climatic conditions. Data permeates designers' interactions with nature. Designers, advisors, critics and developers see nature increasingly through the lens of data, and define nature in its terms.

I have two targets in my sights. The first target is the easy binary that puts technology at odds with nature; second is the trust and power accorded to data. Yes, we can be suspicious of digital technologies, but not because of the myriad changes in practices they require of us. Nor because they encourage 'brain overload', diminish authentic social interaction or distance us from nature. The problem is not devices but data, and the way it filters our perceptions of the world. Occluded by data, it is as if the nature we seek becomes even more elusive and limited. Data is amenable to mechanical manipulation, and there is a lot of it. So, it is easy to see data as a conduit to the unmediated experience of nature. For big-data enthusiasts, data subsumes nature.

I want to unseat the priority of a data-oriented frame. I challenge the importance accorded to *data* in design, and affirm instead its subservience to the rich field of semiotics. Signs, signals and symptoms, and their interpretation, permeate the whole that is nature. Designers, practitioners and consumers can usefully connect to the communicative networks of natural systems. Semiotics contributes much-needed debate about digital technologies and their design, as architects, landscape architects, planners and engineers respond to the challenges affecting landscapes, urban environments and nature. Nature communicates through signs; signs are ubiquitous in nature. But signs are not data. Nor is nature just data. I think that putting data in its place provides a more rounded and rich cultural orientation to the nature–technology relationship. Semiotics also helps us understand nature and health. At least, semiotics and health trade in a common language of signs and symptoms.

Semiotics also provides a means of reintroducing *meaning* into a world view saturated with data. Environmentalist Bill McKibben wrote *The End of Nature*, the first edition of which appeared in 1989.[1] In what sense has nature ended? Assisted by data, human industry manages, controls, scrutinizes and alters natural environments. But nature has also changed its meaning. This proposition provides a further touch point for nature and semiotics. McKibben said of nature that it 'is now a category like the defense budget or the minimum wage, a problem we must work out. This in itself changes its meaning completely, and changes our reaction to it.'[2]

He advances a couple of poignant illustrations of this semiotic shift. The first draws on our aversion to things that we know will end before their time:

> The end of nature probably also makes us reluctant to attach ourselves to its remnants for the same reason that we usually don't choose friends from among the terminally ill.[3]

McKibben claims he chose not to know too much about vegetative die back, and other problems within the natural world affected by humans: 'I like the woods best in winter when it is harder to tell what might be dying.'[4] If he knew what sick trees looked like, he would see them everywhere. Then there is the energy spent on getting out of things, of diminishing commitments rather than sustaining or struggling with them. As marital divorce is so widely accepted, people put their energies into preparing for independence, rather than being interdependent. As if in resignation, McKibben laments: 'There is no future in loving nature.'[5] We will forget how people used to think about fish and other wildlife before the expansion of genetic engineering and industrial-scale farming: 'The loss of memory will be the eternal loss of meaning.'[6]

The natural environment appears vulnerable across many dimensions, not the least its meanings. If we have learnt anything from politicians skilled at manipulating and trading in public opinion, it is that words and meanings really do matter, as does truth. Signs are crucial in understanding the environment and the complex discourses it entails. Semiotics supports this challenge. The stakes have never been higher, considering the threats we face, and how much we depend on what we think of as the natural environment.

ACKNOWLEDGEMENTS

I am grateful for insights into the disciplines of Landscape Architecture, Architecture and environment from colleagues Peter Aspinall, Dennis Dollens, Hannah Drummond, Roxana Bakhshayesh Karam, Michelle Bastian, Rebecca Crowther, Fabrizio Gesuelli, Philip Goulding, Matluba Khan, Dorothea Kalogianni, Asad Khan, Sophia Lycouris, Angus Macdonald, Patricia Macdonald, Panos Mavros, Stella Mygdali, Cristina Nan, Christopher Neale, Dimitra Ntzani, Tolulope Onabolu, Miguel Paredes Maldonado, Andrew Patrizio, Agnes Patuano, Jenny Roe, Graham Shawcross, Katerina Talianni, Neil Thin, Tiago Torres Campos, Catharine Ward Thompson and Dave Wood. Some of the work I reference here draws on the collaborative project 'Mobility, Mood and Place' (EP/K037404/1), supported through the EPSRC/AHRC/SRC/MRC scheme 'Design for well-being: Ageing and mobility in the built environment', which has as its focus well-being and outdoor environments. I am also indebted to the enthusiasm of students in the MSc in Design and Digital Media and the MSc by Research in Digital Media and Culture, as well as teaching colleagues in the Edinburgh School of Architecture and Landscape Architecture in the University of Edinburgh.

LIST OF FIGURES

INTRODUCTION: NATURAL DIGITAL

Digital technology comes off second best when compared with nature. The target of my study is the intriguing relationship between advanced digital technologies and what many choose to contrast it with – variously regarded as a world that is unmediated, authentic and 'natural'. Here, I am thinking of digital technology as a subspecies of 'the artificial' to which the natural is often placed in opposition.[1] Some of us technophiles are excited by the benefits, opportunities and pleasures afforded by digital technologies. To suggest that the digital is somehow at war with 'the natural' goes against our best instincts. But even fans of digital technologies can slip easily into an anti-technological frame.

In a book warning against our obsession with technology Larry Rosen asserts: 'If you are going to use nature as a restorative cure for technologically-induced brain overload, it is best to remove all technology from the scene.'[2] To deride digital technology is a little like insulting your best friend. You know you can get away with it and still accrue the benefits of friendship. Technologies are like friends and family in this respect. Most of us would find it difficult to extricate ourselves from the clothes we wear, our schooling, medical care, transportation and other technological systems, and most of us rely on digital systems for work, leisure and socializing. Perhaps it is this dependence that makes digital communications, systems and devices so easy to write off, especially when compared with things natural. In a book on nature and health Selhub and Logan say, 'instead of stroking the keyboard or rubbing the belly of your smartphone screen, you – and the world – will be better served by petting your dog.'[3] We consumers may be content with the banking systems, airline navigation and myriad communicative networks that we either take for granted or have no knowledge about, but our personal devices bring our dependencies into sharp relief.

One explanation for the preference of nature over artifice resides in narratives of growth and expansion that accompany digital technologies. Technology is on the side of progress. It increases as a conspicuous part of people's lives. Cities expand. Communication networks grow. Technology industries prosper. Narratives about prosperity, progress and the future are on the side of technology. On the other hand nature is on the side of decline. The numbers and varieties of natural species are depleting at a rapid rate. The polar ice caps are receding. Rain forests are being felled. Urban development and industrial-scale agriculture are taking over natural settings. It is increasingly harder to find wilderness, and activists must fight to preserve national parks, green belts, urban parks and open spaces. As far as we bundle up these fading entities as nature, nature is in retreat. It is easy to assume that the more artificial things we fill our world with, the less space there is for nature. Technology crowds out nature. In the David-and-Goliath struggle for survival, we root for the underdog – nature. And like the wise and virtuous David, nature will save us.

Another explanation for the ethical ascent of nature over the digital recognizes that vast technological infrastructures rely on global capital. They also sustain it. As wealth is accumulated by a few, the rest are lulled into the role of acquiescent consumers, buying products they do not need or cannot afford to preserve existing concentrations of capital. Digital systems further exaggerate inequalities, providing new channels for exploitation and inequality to circulate. So far, the escape from capitalism evades us. The best we can do is to maintain a persistent critical stance, and ameliorate the worst of its inequalities. So, critics catalogue and argue particular cases, and keep uncovering iniquitous practices. This is a hugely important task, but such challenges can overwhelm nuanced considerations of the technology. There are also contradictions in the critical stance. Most cultural critics rely on digital infrastructures as they develop their critiques, and are fascinated by the affordances of digital systems, not least how they enable academic critics to communicate and publish. But the positive instrumentalities of digital technologies encounter only grudging advocacy, which takes a resigned or ironic tone. Cultural theorist Benjamin Bratton captures such a mood:

> With significant exceptions, the web has largely been developed through technologies and protocols of British, European, and American origin, with many of the most powerful governmental and economic players still located there. … Its global growth could be read then as the creeping spread of cyber-empire and part of a larger superpower monocultural campaign, starting in Silicon Valley and Washington, DC, and spreading to world capitals like an invasive machinic species. Some European activists, on both the left and the right, describe it this way.[4]

It is scarcely surprising that in contrast to technology, nature has many advocates and very few detractors. Whatever we mean by nature, it is something that is benign, and even good for you. The nineteenth-century essayist Ralph Waldo Emerson (1803–82) described the 'enchantments' of nature as 'medicinal, they sober and heal us'.[5] His friend Henry Thoreau (1817–62) famously retreated from the town to the woods to live the simple life and to get back to the 'indescribable innocence and beneficence of Nature'.[6] Pioneering landscape architect Frederik Law Olmsted (1822–1903) wrote that viewing nature 'employs the mind without fatigue and yet exercises it; tranquilizes it and yet enlivens it'.[7] People readily think that nature and natural settings align with what is good for you. Though people can indulge in unhealthy pastimes in nature settings, we rarely blame nature for bad behaviour. By contrast, unhealthy practices involving computers render technology somehow culpable.

Poets and scholars heap inexhaustible praise on nature. This is scarcely surprising as the love of nature, and some say even its invention, emerged from Romanticism and as a retreat from the excesses of the Industrial Revolution.[8] But of the benefits technologies provide people are less poetical, if not less confident. A press release from a conference run by the British Psychological Society in 2012 stated sensibly that it is in employers' interests 'to encourage their employees to switch their phones off; cut the number of work emails sent out of hours, reduce people's temptation to check their devices'.[9] But

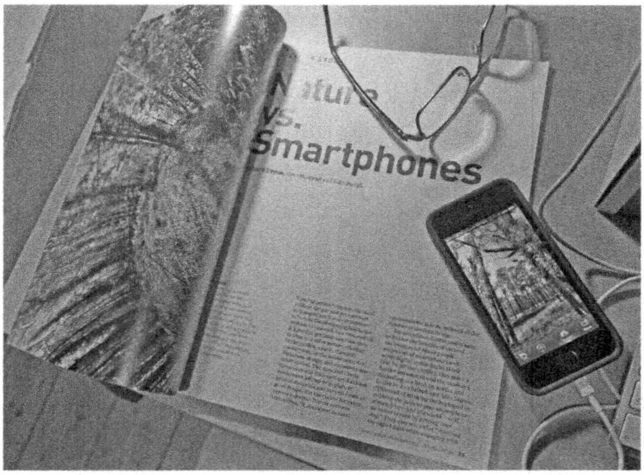

Figure 0.1 Mobile accessories.
Source: author

the press release was picked up eagerly in many online news reports with headlines such as 'Your Smartphone May Be Stressing You Out.'[10] It is an easy jump from advice about moderating your work practices to identifying a malevolent cause – smartphones as the source of stress. Researchers may even bias their evidence gathering to that end.[11]

So, I am tackling the apparent antagonism between the digital and the natural. Human society has long depended on automation, sophisticated transportation systems, mass-media communications and medical technologies. Networked digital technologies entered the dependency chain in the 1950s and have developed at such a rate and yielded such consequences that their adoption still poses fresh challenges well into the twenty-first century. So, some computer users are understandably reluctant and suspicious, and resist the radical technological and social changes that confront them. We use technologies and adopt the practices surrounding them out of necessity. The meagre consolation for this lack of choice is an entitlement to express discomfort and anxiety, to focus life's woes on the machine and automation, and to only seek, collect and report evidence that supports such a case. There is solidarity in identifying a common antagonist, as if sharing grief at something lost. That loss commonly bears the name *nature*.

Digital utopians

But not everyone shares this negative or indifferent outlook. Elsewhere I have examined at length the rose-coloured utopian lens through which many insist on celebrating an inevitable digital future.[12] In this book, I review such techno-romanticism in its renewed aspect as a quest to reproduce or enhance the natural by digital means. As I will show, some architectural scholars advocate for a world that builds on nature's offerings, deploying genetic algorithms and techniques of parametric design.[13] A book on 'hypernatural' architecture advocates for a condition of '*next nature*, a state that transcends current archetypes and provides a more advantageous set of circumstances',[14] heralding such improvement as 'the very aim of evolution itself'.[15]

It is appropriate that I come to the natural–digital pairing from architecture, landscape architecture, planning and the built environment, each managed by disciplines and practices that have over their long histories sought to modify, copy, shape and define nature. Most recently, architecture has responded to the nature–digital challenge by promoting a computer-mediated architecture more closely attuned to nature's forms and processes. This is organic, biomimetic and biophilic architecture.[16] Biology provides potent analogies for architecture in terms of shape, form and process. In his important articulation of the biological analogy in architecture and design, Philip Steadman traces the history of architecture's engagement with things natural, or at least biological.[17] Major players on the analogical stage include D'Arcy Thompson (1860–1948).[18] Among other analogies, Thompson likened engineering structures to the configuration of animal skeletons. He invoked Scotland's Forth Rail Bridge by John Fowler and Benjamin Baker as a prime example. Architects also identify parallels between the classification and evolution of buildings with biological classification systems. Buildings and their components fall within types, genera and families in terms of form and function (e.g. basilica, cruciform-planned church and courtyard house), and there are ample architectural instructional manuals and guidebooks that present these typologies in diagrams as if arrays of biological specimens.[19] There are also 'ecological' parallels. Organisms develop, accommodate and influence their environmental contexts. In analogical fashion, the sourcing of materials, the construction processes and the maintenance regimes of buildings inevitably influence environments, as organisms interact with one another, creating and modifying their ecosystems.

Biological evolution provides a powerful metaphor for design processes.[20] Theorists recruit the evolutionary metaphor to explain the incremental development and improvements of classes of artefacts over time, as makers, manufacturers and designers learn from the mistakes of the past.[21] Think of the evolution of the wheel through its various forms over the centuries. Such apparent evolution is evident not only in craft development over time, but also in highly industrialized, deliberate design improvements, as in developments in glass curtain walls and fixing techniques. Think also of successive

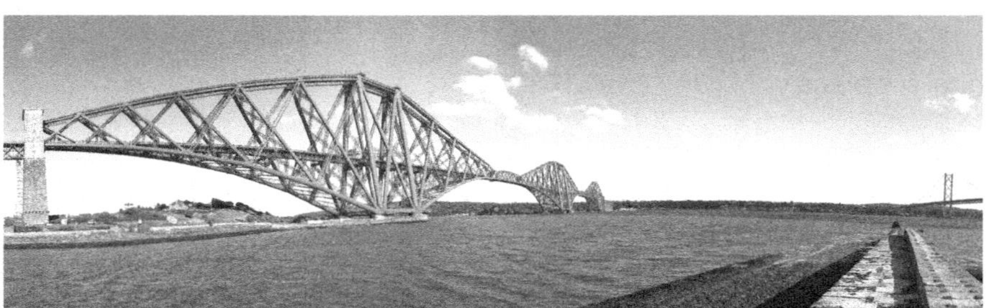

Figure 0.2 Darcy Thompson's dinosaur spine. Forth Rail Bridge, Scotland.
Source: author

improvements in the design of smartphones, and the attendant waste as rival product lines fail in the marketplace. Artefacts exhibit codependencies, and the marketplace encourages something like a Darwinian natural selection among products. Steadman identifies such obvious parallels between design processes and organic processes.

Built environments

Built environment disciplines vary in their orientations to nature. All of them will of course endorse sustainable building, protection of the environment, uses of renewable materials and energy, and other aspects of 'green' construction.[22] Architectural and engineering disciplines seem to identify strongly with biological analogies, perhaps initially through the forms and shapes of the artefacts they produce. In contrast, as indicated by the quote from Olmsted, professionals concerned with landscape are quick to align their relationship with nature in terms of health and well-being. This at least is my own observation from reviewing research from several discipline areas in the UK and within my own university. As further evidence for landscape architecture's affinity with well-being, the UK's Landscape Institute published a position statement declaring: 'Much of the history of landscape architecture can be traced back to the need to create places that were beneficial for people's health and wellbeing.'[23] References to architectural environments that enhance one's health are a relatively recent addition to the architect's lexicon of concerns.[24] Design disciplines aligned with the outdoors, natural settings, horticulture and even farming champion health and well-being, providing further evidence that nature affiliates with beneficence and the good.

Architecture, engineering, planning and landscape architecture are conspicuous consumers of natural resources. As such, these disciplines bring the relationship between the natural and the artificial into focus as they wrestle with energy efficiency, resource depletion, carbon dioxide reduction and other environmental challenges. The built environment is also increasingly designed and mediated through digital technologies, and architecture's custody of a sense of place must expand to deal with the increased presence of virtual, augmented, monitored and data-saturated digital space.

Data flows versus semiotics

Data features prominently in people's thinking about nature and place in the digital age. Data has undoubted utility, with crucial functions not least in furnishing evidence of climate change, pollution and changes in earth systems. But one of my complaints about many of the practices surrounding digital technology is that it subjects everything, nature included, to a common denominator, namely the flow of data. As the philosopher Martin Heidegger said, despite technological achievements, 'the Nature which "stirs and strives," which assails us and enthrals us as landscape, remains hidden.'[25]

I exempt myself from the litany of laments that nature and digital technologies are somehow incompatible. But to recognize the enabling, creative and disruptive potential of digital technologies does not require the technology enthusiast to equate nature with data. Much of my ensuing discussion attempts to address the problems posed by data.

To help position the challenges of big data, nature and environment I turn to the field of *semiotics*, a field already familiar to architects and others concerned with the physical environment, particularly since the publication of the seminal text *Meaning in Architecture* edited by Charles Jencks and George Baird in 1969.[26] The branch of semiotics that is most helpful to the cause of nature considers communication, the transfer of signs, as already belonging to the armoury of nature. Data is just a by-product of the transfer of signs between communicants, agents, elements and the things of nature. That the transmission of signs pervades the processes of nature is among the insights of the American philosopher Charles Sanders Peirce (1839–1914),[27] who represents one of two major strands of semiotic theory. Contrary to Peirce, the semiotic theories of Ferdinand de Saussure (1857–1913) and the structuralists start with human language as the primary means of transmitting and interpreting signs.[28] But Peirce and his followers, notably Thomas Sebeok (1920–2001),[29] think of the communion of signs as existing independently of and before human language. Animals transmit and receive signs, and some of Peirce's followers think that such processes occur at the cellular level[30] and even in geology and nonliving systems.[31]

What is the place of Peirce's thinking in intellectual developments in the twentieth century and beyond? His naturalistic models of communication do not always fit comfortably with advocates of phenomenology and hermeneutics to which many colleagues and I subscribe.[32] Peirce's extensive categorization of signs and dependence on formal logic at times seem unnecessary, particularly to architectural and environmental theorists versed in structuralism, poststructuralism[33] and phenomenology. I shall bring out the relationship between these strands of thought in subsequent chapters. One of my aims is to advance a semiotic view of the nature–digital discourse, above a data-driven view. Semiotics re-invests nature and our experience of it with concepts of meaning in support of a phenomenological orientation.

I develop these themes through eleven chapters, beginning with the common claim that the insistent demands of ubiquitous digital media disrupt how we otherwise 'tune in' to nature. The theme of *attunement* leads me to examine data, and how semiotics presents a more integrated and nuanced understanding of communication in natural and human systems than just data does. The chapters that follow develop the semiotic theme, including comparisons with other points of view, notably phenomenology and hermeneutics. In the process, I canvas mainstream issues in digital technologies as they impinge on place and place making. These issues include those posed by big data, biomimesis, cyborgs, artificial intelligence, video games, robot pets, assistive and therapeutic technologies and digital enchantment. In parallel, I construct a dossier of issues that need greater consideration in the digital world: *attunement*, semiotics, the sublime, monstrosity, melancholy, the trickster function, autochthony, contest, animality, refuge, *musement*, synechism, numinous spaces, and self-reliance. Readers may not yet be familiar with all these terms, nor with their relevance to nature, digital technologies or place – a condition that this book seeks to remedy.

Chapter contents

The first chapter is about *tuning in to nature*. It is a common complaint that digital communications direct the attention of their users elsewhere than the current place. They take digital users out of the moment. We are out of step with the places we inhabit. In this chapter I put this complaint into the wider context of *attunement*, expanding on a theme I developed in previous books.[34] Directing and redirecting attention is a crucial element in what it is to be in tune with one another and with a place. Ideas about *attention* developed within studies of perception, and in turn contribute to how people understand natural environments, as advanced by Peirce's intellectual peer, the psychologist William James (1842–1910).[35] Peirce's naturalist philosophy advocated attunement with nature.[36] After all, evolutionary psychology examines how our forebears survived in hostile landscapes by being able to direct their attention to what mattered. I take seriously the proposition that digital media direct and redirect people's attention in places, and help us to tune in and out of environments.

Having introduced attunement as a major theme of the book, I turn, in Chapter 2 (What Nature), to the kind of nature under consideration in various discourses. I identify the biomimetic strands in architecture, engineering and the built environment, and their legacies. To tune in is also to recognize and interpret *the book of nature*, the subject of the third chapter (The Book of Nature). In this chapter, I begin with data; then I progress to the more revealing lens of semiotics. My conclusion: Whatever nature is, it is not data. In the fourth chapter (Biohacking), I return to the supposed power of data and pursue critically the 'gene as code' metaphor. Though it has undoubted practical benefits, the emphasis on code diminishes the physicality of organic material, especially as we think about the materiality of the physical world and of life.

Human beings manage, frame and copy the things of nature: this forms the subject of the fifth chapter (Reproducing nature). The priority given to data supports the ambition to copy, mimic and reproduce nature by digital means as *parametric* design. Viewed through the lens of semiotics, the reproduction of nature by whatever means illustrates sign systems in operation. Signs inhere within the relationship between humans and the rest of nature as well as in the ways natural systems communicate.

Whereas many critics will accept communicative structures within organic systems, they may be more sceptical about communication within nonliving matter. Do stones speak? In the sixth chapter (The Book of Stones), I examine geosemiotics, progressing to a discussion of earth and the relationship to it of humans as architects of the Anthropocene. In Chapter 6 (Survival of the Fittest), I expand on the theme of *conflict*. Contest, or *agon*, extends to all areas of nature and environment. I examine how communities, urban life, politics, sports, play and video games demonstrate this competitive urge. Humans dispute space, as do other animals. Chapter 7 is about *zoo-space*. I move to consider how non-human animals feature in cognition, categorization, power relations, aesthetic categories and language. This chapter reinforces the contention that natural environments are complicit in networks of signification and cognition, animals are part of that communicative and cognitive scaffolding, and we are part of theirs.

The human animal craves narratives indicating a positive relationship with its environment. In Chapter 8, I turn to *refuge*, and examine the positive values that people attach to what they think of as natural places – beautiful, healthy, restorative, interesting, engaging, protective and fulfilling. But there are other kinds of places. Chapter 10 focuses on *numinous places*, in which 'the book of nature' is less transparent to our reading practices. It is as if certain places emit direct parapsychological connection between people and places. Fantasy, science fiction and predictions about the opportunities provided by digital systems feed off some primitive urge in all of us to identify immediate connections with nature. For some, digital technology offers a means of transcending the physical limits of space and society. But what happens when the machine stops? In the eleventh chapter (Nature Unplugged), I examine how we humans try to be independent from mechanization and artifice. We recruit digital technology to define and mark out the autonomous, independent individual. To conclude my main narrative, I return to Peirce who was operating within and responding to a climate informed by a popular romantic naturalism and the desire to escape the machine and get back to the woods,[37] an impossibility, inevitably unrealized.

Method of inquiry

This book is born from my own practical inquiries, and that of other technology users and designers with whom I work. Some of us have undertaken several empirical projects that study how people respond to changes in outdoor environments, in particular as they walk through busy urban streets and parklands. We have gathered biometric data from participants by means of portable electroencephalography (EEG) equipment and interviews. That work is reported elsewhere[38] but informs this study, not least in that it helps me test the strengths and limits of biometric data.

Architecture, urban design and landscape architecture have long positioned themselves within a multidisciplinary frame. Part of the architectural legacy from which I draw considers all sources as potential contributors to a case. I canvas many sources, including philosophy, biology, geography, social science, cultural theory and the arts. Architecture at times assembles its disparate parts into a whole. The parts interconnect in unexpected ways. I may appear to digress at certain points in this book. But I keep returning to the book's three main actors: place, digital technologies and nature.

As with previous publications, this book arises during a period of experimentation with public writing.[39] I have rehearsed some of the ideas here in blog format, in the shape of a public notebook, then reshaped and assembled them as a coherent narrative that positions network nature in the wider semiological sphere.

CHAPTER 1
TUNING IN TO NATURE

Tuning in is a popular metaphor in the electronic age. Before the introduction of push-button controls, listeners would turn a dial of their radio and lock on to a frequency band to receive a signal. To tune in is to pick up on a frequency, to filter out all the other frequencies and home in on a specific band width, for example, 92.5–96.1 megahertz.[1] To tune in to a channel is to pick up a signal, which is in turn a carrier for other signals, including speech, sounds and music. In the case of broadband media, the signal conveys moving images and interactive elements, buttons, fields and haptic interactions. These media in turn circulate words, meanings, actions and ideas among broadcasters and receptive audiences. 'Signal' is a helpful term as it resonates with the idea of tuning and *attunement* – tuning in to a signal – but 'signal' implies something raw, abstract and unmediated. Scholars in the humanities prefer to use the more general term 'sign', from which 'signal' is derived, as an indicator of what it is that is circulated when people and things communicate. The idea of the *sign* does not presume some base-level primitive component of communication, such as a frequency, a vibration, perturbation or data, and leaves open the role of the interpreter, the context of interpretation, including the communicative environment and the culture in which the sign operates.

The study of the communication of signs is *semiotics*, as advanced by Ferdinand de Saussure,[2] C. K. Ogden, I. A. Richards,[3] Roman Jakobson and Fredric Jameson.[4] But here I will mostly reference the work of Charles Sanders Peirce,[5] William James[6] and John Dewey[7] of the American school of pragmatism, mainly because of their attempts to shed light on our experience of the natural world. Their philosophy has in turn informed more recent theorists who study the human–nature relationship. Writing in the latter part of the nineteenth and early part of the twentieth centuries, these pragmatists also presented a world view before the influence of digital media, before natural phenomena succumbed to digital data.

Attunement

The tricky part in tuning a radio dial is to listen to the sounds, distortions and other clues and signs indicating that you are close to the target channel. To expand the metaphor, the key means of *tuning in* is to pick up the signs and cues of a place, deliberately, though sometimes without conscious attention. *Tuning in to nature* is one of the ways gardeners, outdoor adventurers, environmentalists and citizens strolling in the countryside describe their relationship to the world of nature. Peirce's 'naturalism' advocated *attunement* with

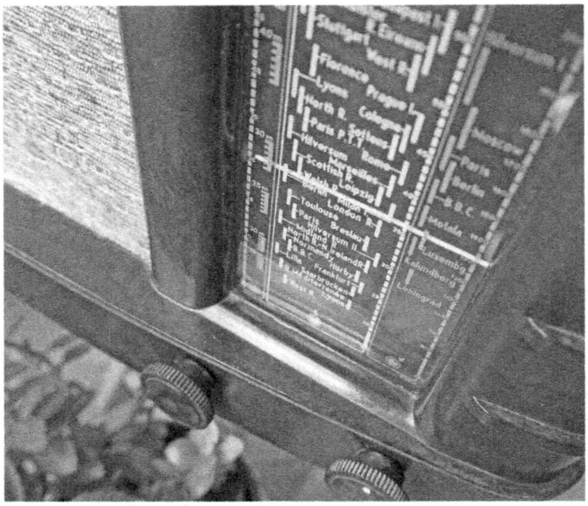

Figure 1.1 Radio tuning dial. 1940s Philips 206a Bakelite Art Deco valve table radio.
Source: author.

nature,[8] and so is an appropriate source for advancing our understanding of the nature–culture relationship.[9]

The concept of tuning extends beyond radio transmission, and of course predates it. To tune in is to adjust to any kind of vibration, regular oscillation or repeating process. The vibrations that constitute sound provide an obvious context for tuning. Musicians tune their instruments, and auto-mechanics tune engines. Moving into the culturally nuanced realms of daily living, we see that workers, citizens and visitors tune in to the everyday rhythms of the city.[10] Natural environments are replete with rhythms over varying time scales: the cycles of the seasons, diurnal rhythms, waves lapping on the sea shore, and the beat of animal feet, wings, calls and body gestures. In a general way, for human beings to tune in to nature is to pick up its rhythms, to adapt and synchronize in some way. People adjust their rhythms to engage with the complex overlapping rhythms of the worlds they inhabit, including the world of their own bodies. No doubt, there are many ways to achieve this synchrony, including mimicry, copying what other people do in that setting, and even copying the sounds, sights and behaviours of the environment. Some people are adept at surrendering to certain situations, as if a piece of driftwood carried along on the currents, waves and tides. I focus on signs, but we should be open to the possibility that there are many aspects of our interaction with the environment. I will consider these in Chapter 10, which deals with numinous spaces. One may also tune out and resist the rhythms of a place. But to be out of step, deliberately, requires some awareness of the rhythms of the environment one is in. It takes effort to go against the flows and rhythms of life, and you have to be tuned in to tune out.

Tristan Gooley's book entitled *The Walker's Guide to Outdoor Clues and Signs* provides further illustration of how people can tune in to nature's signs and rhythms. We learn that darker clouds presage rain, reeds growing by the side of a lake indicate areas where there are fish, and a moon less than half full means high tide for that day will be below the maximum high-water line. Such signs depend on the cycles of the seasons, weather, position in the landscape and the ways in which plants and animals adapt to locational variation. According to Gooley, 'All cycles are interrelated, and once we are familiar with one part of these clocks and calendars, we can use it to read others.'[11] To be attuned to nature is to be attentive to its signs and cycles.

Paying attention

Attention is key in identifying signs, and hence our reading of the natural world.[12] After all, survival in the world requires sign-reading skills. Think of the human animal struggling for survival before there was agriculture, structured community life, domesticity and cities. Without the ability to attend to the world around us, to pick up its signs, our ancestors (and we) would be subject to an ocean of sensations that drowned out effective action. You can also think of attention as *interest*. Attending to the signs, objects, situations, problems or tasks at hand engages us and allows us to cope, survive and thrive.

Attention often requires work, that is, concentration. That our attention may wander from the task at hand is not altogether detrimental. By various accounts,[13] if we attended exclusively to what mattered at that moment then we would not be so alert to other important signs competing for attention. While attending, with fascination, to the flowers in a meadow we may not notice the sign of an approaching predator rustling through the grass.

The idea of the sign relates to perception and sensation, that is, seeing, hearing, touching and in other ways sensing the world around us. The philosopher William James (1842–1910) said that 'we notice only those sensations which are signs to us of *things*',[14] elaborating that such things are those that present to us as of practical interest. James wrote less about signs than the processes of attention and different categories of attention. Attention, and attending to signs, is important in developing our relationships with nature. He said there is attention that is aroused by some sensation (hearing signs such as an unusual bird song, or observing a shooting star). Attention can be intellectual, as when struggling with a maths problem – think of mathematical symbols as signs – or deciding whether to take an umbrella to the park, which involves reading the weather signs, and conjuring up mental images (signs) of getting wet. Then there is attention that is immediate, as when the object of attention is interesting 'in itself', as in the case of a table of appetizing food, a treacherous waterfall, bleeding wound or other signs of direct benefits or threat to the human organism.[15] According to James, there is also *derived attention* that associates with rich cultural meanings. A tapping sound is mildly arresting as a sign that calls us to action, but in sentimental vein, he reminds us that 'when it is a signal, as that of a lover on the window-pane, it will hardly go unperceived'.[16]

James identifies voluntary attention that is hard to maintain, as when concentrating on an important, tedious or demanding task, such as reading an instruction manual, or writing an essay about a topic that does not interest you. It is easy to get distracted from such tasks by something easier to manage, such as checking social media updates, or

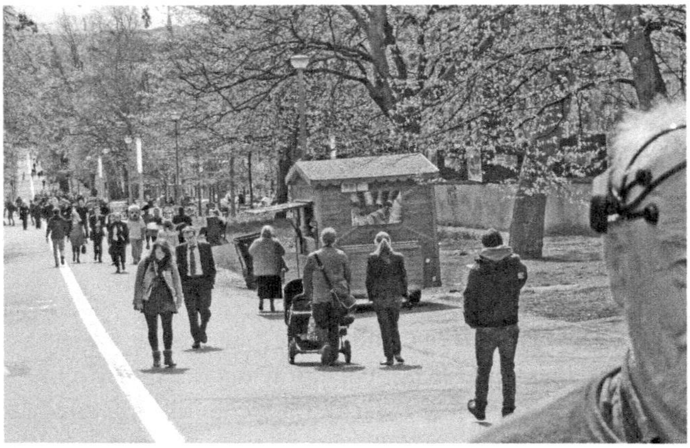

Figure 1.2 An exercise in attention: the bear in the park. Meadows Walk Edinburgh. *Source*: author.

leafing through the jokes on a desk calendar. When struggling to write an essay about something we know little about or to create an inventory of expenses claims or other intensive activities that require us to monitor, record and manipulate recalcitrant signs, our attention diverts easily to signs that are extraneous, not least the insistent calls for attention from email, text messages, and social media updates.

On the subject of distraction, we can be blind to potential signs around us. The publicity brochure for a research project[17] I am involved in includes a photograph of people walking through a park. There is someone in the foreground wearing an unusual head-mounted EEG apparatus. Everyone notices that. But few of the people who have seen the picture notice the person in the distance wearing a bear costume.[18] That usually goes unnoticed, even by the photographer.

Embodiment

Lest we think that such attention to signs is purely cognitive, or all in the head, James explains these attentional behaviours as derived from our condition as embodied beings. He says:

> When we look or listen we accommodate our eyes and ears involuntarily, and we turn our head and body as well; when we taste or smell we adjust the tongue, lips and respiration to the object; in feeling a surface we move the palpatory organ in a suitable way.[19]

Some cognitive researchers like to think of attention as something happening in the organ of the brain. Writing in the very early days of neuroscience, James was convinced that bodily exertion precedes cognitive effort.

To pay attention is at root a bodily activity. Teachers encourage pupils to sit up and take notice.[20] Soldiers on parade stand to attention. Carers touch, gaze, support, coax and engage with the body of the other.[21] That is what it is to attend, and to be in attendance. To care is to look, listen and to pick up on signs of discomfort and distress that are current, potential and imagined. Empathy is after all an immediate response to the bodily actions of others.[22] For James, positive and negative bodily actions result in 'a more or less massive organic feeling that attention is going on'.[23]

Design deals in functions related to the body. Theorists such as Geoffrey Broadbent[24] and Donald Preziosi[25] have examined the relationships between signs and functions in architecture, and social philosopher Michel Foucault[26] foregrounded the role of the body among design thinkers. He examined how spatial arrangements codify body and power relationships, and how these are evident at different stages in history: from how desks are set up in classrooms to the way hospital wards confine patients. It is fair to say that body, sign and function elide. They are key aspects of any encounter with the world.

Signs and emotions

Signs also carry emotional entailments. The psychologist Wilhelm Wundt (1832–1920) incorporated *attention* into a simple emotional scale.[27] A human emotion registers as an intensity of feeling: very strong to very weak; and a pleasure scale that ranges from highly pleasurable to very unpleasant.[28] Wundt introduced a third dimension spanning a spectrum from 'strain' to 'relaxation' that accords with the idea of *attention*.[29] It matters for the emotional valence of an experience whether you are attending to the experience or if it is somehow in the background of your awareness.

There is something immediate about our appropriation of signs, and of the emotions they entail, an insight that will later on assist in thickening the link between semiotics and nature.[30] The phenomenologist Edmund Husserl asserts as much. The world is not a catalogue of 'facts and affairs', but for the person who is conscious and aware, the world appears as 'a world of values, a world of goods, a practical world'.[31] His immediate experience is of things 'beautiful or ugly, agreeable or disagreeable, pleasant or unpleasant'.[32] This is what it is to attend practically to objects and signs in the world: the path, the view, an oak, or the meadow apprehended as steep, breath-taking, solid or restful.[33] This alignment of signs and emotional responses will later support my case for a semiotics of nature.

Intention

The information model of Claude Shannon and William Weaver that drives many digitally oriented communication models assumes a sender who has something in mind to be communicated to someone else.[34] The sender packs the message into a signal (e.g. words) that is then communicated to someone else who unpacks the intended meaning. Many scholars from semiotics and structuralism have challenged this model.[35] The metaphor assumes that meanings get packed into texts, that are then transmitted; then the receiver unpacks the message. Meanings pass through such communicative conduits. Shannon and Weaver's information theory was also preempted several decades earlier by Peirce's pragmatic account of human intentions.[36] Intentions are not packed into the words, but reside in the shared communicative practices, and the sense of responsibility and accountability, to which the speaker is prepared to subscribe. After all, many of us would rather deny an intention than claim one. Life is too random to be certain about who intended what.

The problematic of climate change provides a potent test case for intentionality. According to psychologists Ezra Markowitz and Azim Shariff:

> Although climate change is the direct result of intentional, goal-directed behaviour (for example, the use of energy to provide all the trappings of modern life), it is probably perceived by many individuals as an unintentional, if unfortunate, side effect of such actions.[37]

Intentionality has powerful connotations in contemporary nature narratives: 'Unintentionally caused harms are judged less harshly than equally severe but intentionally caused ones.'[38] Intentions and their absences run deep in human interactions with nature, in terms of signs, sources, causes and agency.[39] Husserl wrote extensively on *intentionality*,[40] and asserted that conscious perception involves 'being turned towards an object',[41] a theme that Heidegger would later explore in *Being and Time*.[42] Being so turned towards an object also resonates with the bodily practices of attending to signs in the world.

Prospect

Caspar David Friedrich's iconic painting 'Wanderer above the Sea of Fog' (1818) depicts the aesthete attending to the broad and distant mountain view, a common theme in romantic paintings.[43] What is it to attend to nature? In his book on outdoor clues and signs, Tristan Gooley says that 'a good view of a landscape is not just pretty, it's also a rich source of information'.[44] He then outlines some of the advantages of being high up in the landscape, such as being able to see other high spots (trig points) that are in turn signs that aid navigation. People attend to broad sweeping views of spectacular landscapes, and pay a lot of money for a restaurant table, hotel room or apartment with a view of the mountains, a lake or the wild, untrammelled countryside. The geographer Jay Appleton (1919–2015) advocated that people prefer views, scenes, paintings, and by implication, landscapes, in which there is an element of both prospect and of refuge.[45] In my semiotic frame of inquiry, we could say we prefer landscapes that contain signs of prospect and refuge, that is, tall towers, peaks and ridges that signal the possibility of further prospect, and groves of trees, the shadow of a cave mouth or a hut in which we could attain refuge. The theory is that we are programmed biologically to seek out places where we have a view, to identify the approach of danger (threats from predators), which at the same time provide somewhere to hide. Sheltering in a copse of trees on the side of a mountain might provide this, or its simulation as a view from a balcony. The hunter and the hunted are drawn to settings in which they see an opportunity for prospect: the sight of cliffs or towers in the landscape, and the sight of places to shelter.

Appleton later qualified his hypothesis in favour of movement rather than the static view:

> Very often the balance that can be achieved from serial vision, involving the successive experiences of exposure to strongly contrasting landscape types, strong prospect then strong refuge, is more potent than that which comes from trying to achieve a balance all at once.[46]

It is as part of a sequence that the experiences of prospects, refuges and hazards have their effect. Architects such as Gordon Cullen (1914–94) also knew the importance of movement in the world of signs.[47] The sight of something unexpected, dramatic or spectacular awakens our senses, especially when on the move. Gordon Cullen produced alluring drawings of 'cityscapes' in the 1960s, which show the progress of a pedestrian through a historic town. Cullen describes one such sequence:

> The even progress of travel is illuminated by a series of sudden contrasts and so impact is made on the eye, bringing the plan to life (like nudging a man who is going to sleep in church).[48]

I like the metaphor of the nudge as it foregrounds the role of the sign in such transitions, and how signs arouse people to attention. Such encounters also keep us moving. Cullen speaks of a journey through the town of Oxford, which reveals 'the unfolding drama of solid geometry'. He adds: 'This is the unfolding of a mystery, the sense that as you press on more is revealed.'[49] As in the case of Gooley's *Walker's Guide*, the ability to pick up clues and signs in the landscape, and transitions between such signs, is motivated by the need to keep on the move, to track, to navigate and to journey as in the mobile life of a trekker, nomad or urban flâneur.

Around the same time as Cullen wrote his work on moving through towns, Donald Appleyard, Kevin Lynch and John R. Myer were studying views encountered while driving at high speed along freeways: 'In periods of wide scanning, attention regularly returns to the road itself.' They add:

> The only exceptions to this rule occur in those brief periods where the observer passes some important barrier and, being anxious to reorient himself, surveys a new landscape.[50]

This 'visual revelation' makes a sudden demand on the driver's attention. Here is how they describe travelling across Manhattan:

> Can any driver be ignorant of his passage under the George Washington Bridge, or his entrance into the Holland Tunnel? These are all opportunities for visual emphasis that will claim attention despite a normal state of distraction. The silhouette of an overpass, the texture of a retaining wall, the shape of a bridge column, guard rail or lamp standard are important events.[51]

The experience is prompted by transition from a confined route, to one of prospect: 'Another striking feature is the importance of objects in axial view as the road comes out of confinement: the water tower, the Seagram's sign, the Custom House tower.'[52] The impressive prospect is also of other opportunities for prospect, iconic landmarks, with towers among them, even if inaccessible from the road. Landscape architect Sylvia Crowe (1901–97) provided similar approving accounts of views of power stations and cooling towers in the countryside when suitably sited and landscaped.[53]

I have already alluded to the emotional entailments of signs of various kinds. Appleyard, Lynch and Myer describe a driver's experience that is accompanied by a kind of melancholy:

> On rising to the crest, he expects some announcement: a view of the city, a new landmark, or a more vivid view of a previous goal. But the confinement of the dip is succeeded only by a bland and featureless horizon. Confinement and hiding, without equivalent visual intensity upon release, is a disappointment to the observer.[54]

When anticipations are not fulfilled then disappointment may follow. It seems to me that signs in the landscape, views, prospects and places are in the company of a range of emotions including awe, pleasure, delight, recovery, repose and safety as well as disappointment, anxiety, danger and melancholy. But all this takes place in the context of the movement of and through signs and their attendant contrasts, either actual or implied. The sudden impact of such signs also draws our attention. The effect may be fleeting, when as a hotel guest you walk on to the balcony for the first time and remark on the view. It grabs your attention though you pay it less regard subsequently, unless showing it off to others, or you are moved to another room. Such is our experience of place, amplified and thickened in the case of our attention in and to the world of signs in the environment.

The aha moment

You drive, cycle or walk through heavy traffic, road works and urban blight. You turn a corner, pass over a ridge, emerge from an urban canyon, and something else is revealed – a wide vista, an ocean view, a copse of beech in the low winter sun. You are taken aback momentarily, and remark favourably on the spectacle to anyone travelling with you: 'A turn in the drive; and suddenly a new and secret landscape opened before us,'[55] reported Charles Ryder in Evelyn Waugh's *Brideshead Revisited*. I recall something similar when I once approached Blenheim Palace from the narrow streets in the village of Woodstock in

England. Turning the corner from the Market Square and Park Street, you look through a gate at a landscape of an altogether different scale. It is a magic moment: the grassland, bridge, lake, clusters of trees and the palace. In *The Lion the Witch and the Wardrobe*,[56] C. S. Lewis describes how Lucy finds herself secreted in a wardrobe full of fur coats. She ventures deeper into the wardrobe and the fur coats become fir trees, the white mothballs turn into snow, and she is in Narnia. These are 'aha' moments. You may have progressed from the ugly to the beautiful, and from the mundane to the magical. But more importantly, you have transitioned from a world that is congested and difficult to one that is clear and coherent. In the twinkling of an eye, and for a moment at least, the world makes sense – effortlessly. This is also a semiotic moment. To move from a position of confusion to one of legibility is to move from a place populated by conflicting signs to one in which there is a clear sense that something is being communicated. The landscape delivers a simple message, or the signs connect, at least for a moment.[57] Our images of nature often invoke such clarity. This is how the world is meant to be.

The Epicurean philosophers affirmed that nature provides such moments. The trials of war, the perils of life at sea, and the luxuries of city living diminish 'when men recline in company on the soft grass by a running stream under the branches of a tall tree and refresh their bodies pleasurably at small expense', said Lucretius (99–55 BCE).[58] And from the pen of Blaise Pascal (1623–62) we read:

> Let man then contemplate the whole of nature in her full and lofty majesty, let him turn his gaze away from the lowly objects around him; let him behold the dazzling light set like an eternal lamp to light up the universe, let him see the earth as a mere speck compared with the vast orbit described by this star.[59]

Then follows an account of the perspective gained as we contemplate the infinite: the experience of the aesthete in the face of nature, who attests to a clarity of vision, a renewed sense of perspective, restored balance, the uncluttering of thought that comes with time spent in the midst of nature. This kind of arrested attention follows a period of effort, and is accompanied by relief that it was worthwhile.

Moments of clarity in the face of nature parallel the *aha* moment in which a solution reveals itself in problem solving. Take camping as an example. I open my camping equipment to discover that one of the tent poles is missing. I debate whether it is worth returning home to retrieve the pole, I should try to borrow one from elsewhere or should I experiment to see if the tent will be secure without the pole. None of these seem practical. Then the solution suddenly occurs to me – position my car and tie one side of the tent to the roof rack for support. An article by empirical psychologists Sascha Topolinski and Rolf Reber summarizes four aspects of such an *aha* moment, which they also describe as the moment of 'insight' for someone trying to solve a problem.[60] First, the insight happens suddenly. Second, it is an easy transition appearing to require little cognitive effort. Third, the outcome is pleasurable. I am pleased with myself for alighting on such a simple solution. Fourth, I am briefly convinced of the truth of the solution, even before we have tested it. I even feel compelled to defend the solution against critics and detractors. After all, using the car to support one side of the tent seems like a good idea until someone wants to drive off to buy milk. Such problem solving involves the manipulation of signs as propositions in language, with elements such as *tent, pole, rope, support, wind, journey, home, memory,* and *error*. However elusive, there is logic to such a process, which I elaborate further in Chapter 7 where I consider Peirce's concept of abduction and the semiotics of play.

Transitions

How does such problem solving relate to landscape? Travelling through landscapes involves transitions from one condition to another – changes in the sign landscape (signscape) marked by openings and turnings around so many corners. As I explored in *The Tuning of Place*,[61] pedestrians pass over the thresholds of doorways, cross over from the footpath to the road, from inside a building to the street, from one precinct to another, from an orientation at right angles to one parallel to the street – encountering changes in spatial configuration, geometry, surfaces, materials and levels. Not all threshold encounters are visible as lines, corners or marks in space, though travellers may detect their effects, which is to say their *signs* in the environment, sonic, tactile and olfactory. There are zones of microclimatic variation, areas of light and dark, transitions into and out of the shadows. Nature affords many such transitions: a move from the forest to an open space, light to dark, the valley to the prospect over the dunes, wet areas to dry ground, loud cataracts to quiet meadows, acrid to sweet breezes, and exposed beaches to sheltered coves.

Much lab-based research that attempts to monitor human responses to stimuli focuses on transitional events – the changes in neural patterns or skin conductance that occur when there is a change in the stimulus: for example, a change in visual stimulus, a musical tone, a slip in the rhythm or an adjustment in key. Such changes register as peaks in a person's arousal, or other bodily states.[62] Something similar occurs in a traveller's transition through the environment. It is the *aha* moment, the shock, the discovery, the shift in environmental stimulus, the threshold crossing and change in the signscape that registers as a blip in emotional key. The aware walker turns a corner, registers a significant change in emotional register, tunes in to a new mood, shifts his/her attention.

With colleagues, I have explored several scenarios, taking EEG readings while people walk through different environments. The most dramatic reading occurred when we organized a walk through a pedestrian-and-cycle tunnel in Holyrood Park in Edinburgh. The route passes through a disused rail tunnel about 200 metres in length. The neural response of our walker was most pronounced as she transitioned from the tunnel into daylight at the end.[63] As is the case in storytelling, a skilful architect, landscape architect or planner might modulate alternations between places that frustrate and those that present prospect and resolution.

Frustration looms large in people's thinking about artificial spaces and their technologies. Cumbersome fingers swiping across overly sensitive tablet screens shut

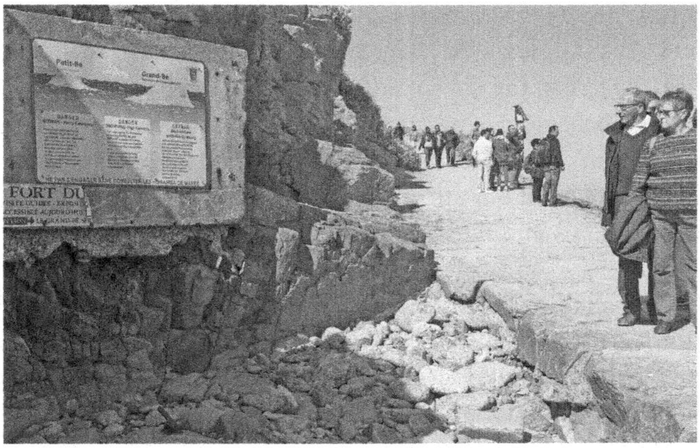

Figure 1.3 Warning sign in coastal landscape. Le Grand Bé, St Malo, France.
Source: author.

down the mail app before the message is finished, windows appear or disappear as if at random, and there are all those irrelevant pop-up messages, obtuse instructions and uninvited features. The physical environment invokes similar frustrations. As it happens, 'frustration' is one of the parameters picked up by the Emotiv Epoc EEG monitoring headset[64] used in the tunnel walk I just described. The technology is designed to monitor a game player's responses to events in a video game, though we adopted the category in monitoring people's responses to outdoor environments.[65] Frustration is a feeling that arises when your ambitions towards a goal are blocked (e.g. there's no tent pole or milk for the tea).[66] I am fond of Marc Augé's characterization of non-place,[67] ably illustrated in the case of the confusion of signs around a fire escape door in a building. It is a frustrating place to be, but there is also a suspicion that I should not really be in that place at all, let alone pass through the door. Similar signs (notices) appear in the countryside, particularly in places that entail hazards. Experiences of natural environments, with and without such artifice, are replete with transitions from threatening to safe, restricted to open, and confusing to clear in nature's rich ecology of signs

Habituation

To be in tune with nature implies some accord with the patterns and cycles of nature's sign ecology. Patterns that are once established require no attention, unless you want to change them. William James characterized the inertia evident in people's behaviour patterns: 'Habit is thus the enormous fly-wheel of society, its most precious conservative agent.'[68] James also draws on nature metaphors to illustrate his discussion of habit. He considers how a pool of water eventually finds its way to an outlet. From then on, the water drains more easily, following the same channel, as the channel gets deeper and impediments to the flow are worn away. So too for the human cognitive and motor systems: it is as if repetition of a task makes certain flows along the nervous system smoother, such that the task becomes a habit. Elsewhere I have considered habits, habituation and habitats in conjunction with the role of the mass media and digital media that reinforce and disrupt habitual patterns of behaviour and practice.[69] Habits involve signs. 'Habit diminishes the conscious attention with which our acts are performed,' said James.[70] In fact, while doing something out of habit our attention can be directed elsewhere. Hence most of us who drive a car frequently and safely do so while attending to something other than the driving – the car radio, a conversation, the scenery or daydreams. James suggests that habit inheres within all things, nature included:

> The laws of Nature are nothing but the immutable habits which the different elementary sorts of matter follow in their actions and reactions upon one another.[71]

As a pragmatist, James was keen to unite psychological phenomena with the rest of the natural world. His follower John Dewey added further support, asserting that nature is experience all the way through, experience 'tunnels in all directions'.[72]

This discussion of signs and their consort with attention are relevant here not least as many of us suspect advertisers deploy discrete tactics that monetize user attention as we browse web pages and follow social media feeds. We give up click and scroll data to Google and others. Advertising revenue accrues via optimized animations and screen displays, which we scarcely notice, but that influence us to buy or at least orient ourselves to consumption and calculative political messaging. In the worst case, targeted messaging is used to confuse.[73] Nature and environment are casualties as public opinion turns on the interests of hidden persuaders, and their power structures and profits.[74]

Of greater interest to my case here, signs are also relevant in the complaint that digital devices and communications distract us from the signs of nature, or at least things that are authentic in our sociability, and in our places. Such is social psychologist Sherry Turkle's case that people seem more interested in their smartphones than what is going on around them.[75] The claim here is that digital communications direct the attention of their users elsewhere than the current place. Such devices take digital users out of the moment. We are out of step with the place. Our habitual practices are disturbed; we adopt unsociable habits and 'addictions'. In this chapter, I tried to put this complaint into a wider context. Directing and redirecting attention is a crucial element in what it is to be in tune with one another and in a place. I have endeavoured to approach nature from the point of view of experience and attention. This serves to introduce biosemiotics and geosemiotics, which will be discussed in later chapters. Next, I will review other key themes that form the background to my study of nature, place and digital technologies.

CHAPTER 2
WHAT NATURE?

I have already referred to those digital technologies with which many readers are no doubt familiar, such as networked and ubiquitous smartphones, tablets, streamed media channels, video games, digital imagery and applications in the workplace. But in examining the idea of *network nature,* I also think of inconspicuous sensing technologies, networks, business-to-business applications, and applications of computing in remote sensing, mapping and data-intensive science. Martin Heidegger asserted that technology is 'a way of thinking'.[1] Technology and social systems work together as 'socio-technical systems'.[2] Elsewhere I have addressed digital technologies themed around overt techno-rationalism, techno-romanticism, e-commerce, sound, attunement, mood and emotion.[3] My focus on technology typically draws on phenomenology and hermeneutics.

Attentive readers will have noticed already my varied uses of the term 'nature'. I have alluded to the natural as if that is what is left once you take away technology. The natural also presents as a world available to direct experience unmediated by technology. The natural has a place in the narratives we tell about the world and ourselves. I have also alluded to nature's association with what is good and right, and its association with health and well-being. There have been movements, some with political and economic consequences that rally behind the banner of nature, not least the various environmental, peace, counter-cultural and democracy movements.[4] As philosopher Judith Butler asserts, nature is not something that comes before culture as 'a politically neutral surface on which culture acts'.[5] According to many commentators nature politics have informed Silicon Valley culture, and hence the world of digital technologies.[6]

Nature also features in many narratives of radical transformation and crisis: 'Our relationship with nature is changing' and 'Nature changes along with us', affirms the online book *Next Nature.*[7] 'Nature' written with a capital 'N' harks back to a convention that lingers from the personification and deification of nature, as in Thoreau's book, *Walden.* That convention helps distinguish the word from more prosaic uses, such as where Thoreau says, 'I am by nature a Pythagorean.'[8] Here, to identify the *nature* of a person or thing is to acknowledge their tendency to follow an inclination, custom or habit. Concepts of nature also carry import when connected with locations, that is, natural environments, settings and places. As habitats, natural places are places of habits,[9] an insight that entitles those who are concerned with the physical environment, such as architects and geographers, to claim nature as one of their categories.

In seeking clarity on the definition of 'nature', it is worth paying attention to the Pragmatists' approach to meaning. Among C. S. Peirce's elaborate definitions and terminological invention is the affirmation at a basic level: 'A word has meaning for us in so far as we are able to make use of it.'[10] This practical stance to meaning and the meaning of nature pervades the social sciences, and has been given ample treatment by Phil Macnaghten and John Urry in their book *Contested Natures*. They provide a stark account of the diverse ways in which nature presents to us. They want to show that a single uncontested nature does not exist independently of the way people study, use, debate and negotiate it.

Macnaghten and Urry hold a position contrary to three conventional ways that scholars think of nature. The first is *environmental realism*, which asserts that nature exists independently of our reflection, and is best probed by scientific analysis: 'Social practices play a minor role in any such analysis since the realities which derive from scientific inquiry are held to transcend the more superficial and transitory patterns of everyday life.'[11] The second is *environmental idealism*, which assumes that there are underlying and stable values that define natural places. I would position this with those who believe in the spirit of place, a romantic conception of the natural world, common enough in architectural discourse.[12] Third is *environmental instrumentalism*, which draws on cost-benefit analyses to arbitrate on the value of nature and natural environments. Nature is what we must take care of for future generations and for the greater good.

In contrast to these positions Macnaghten and Urry's is sociological, cultural and pragmatic, asserting the crucial place of discursively contested social practices. They accord importance to the way people talk about nature in everyday language. They think it is also significant how people sense and experience nature bodily; how people comport, utilize and talk about bodies (their own and other's) that invoke concepts of nature and the natural. Macnaghten and Urry also attach importance to spatial relationships and the way space is defined, managed or left alone, especially through concepts of the relationship between the local and the global. Time is also important in their conception of nature, particularly as those in authority construct the past and future, and plan for uncertainty. Finally, nature discourses are structured around frameworks for organizing human activity, involving risk, agency and trust. Macnaghten and Urry contend therefore that

> there is no singular "nature" as such, only a diversity of contested natures; and that each such nature is constituted through a variety of socio-cultural processes from which such natures cannot be plausibly separated.[13]

I accord some space to this précis not least as I agree with their position, and attempt to flesh it out with a broader range of cultural evidence and an emphasis on design and the digital.

They wrote their book in the 1990s. I think networked digital technologies now reinforce their themes in interesting ways. On the theme of discursive social practices, people talk about, represent and reflect on the natural through a diverse range of everyday digital media, such as digital photography, video, ad hoc travelogues in social media, and the use of navigation tools. I like to think of nature *reconfigured* via these media. Digitally moderated encounters with the natural represent bodies in relation to nature in some new ways. Think of how people put themselves in the picture when filming nature scenes. With a smartphone, you can be in a local situation, but at the same time access information, post a commentary on Facebook, and in other ways play around with, test and subvert the spatiality of where you are. On the theme of space and time, as well as obviating some of the time constraints of travel and communication, networked digital devices influence recollection, anticipation and planning. Via itineraries and other organizational tools ubiquitous digital media inform the way we order the day, the outing, as well as recording, monitoring, and way finding. On the theme of activity framing, I like to think that digital networks are complicit in configuring and reconfiguring sets of relationships that constitute the natural. Imagine tourists or outdoor sports enthusiasts recording videos of a walk in the countryside using their smartphone or paragliding with a 360-degree or wearable GoPro video recorder. Such representations throw up differences. The video is nothing like being there, but might just emphasize aspects of the experience of nature that otherwise go unnoticed – the horizon, risk, equipment, orientation, directionality, framing, and vertigo – and occlude others – the sociability of the experience, waiting, preparation, duration, smell and the wind.

As suggested in the previous chapter, it is tempting to think of the natural world as a casualty of sophisticated communications technologies. Everywhere, always-on networked phones and computers diminish spatial demarcation, and threaten the uniqueness and placeness of natural environments. Macnaghten and Urry draw on various sources to argue the converse: 'As spatial barriers diminish, so we become more sensitized to what different places in the world actually contain.'[14] As evidence of the value attached to the idea of nature they also show the commercial use of the 'nature' label as an asset that helps sell goods:

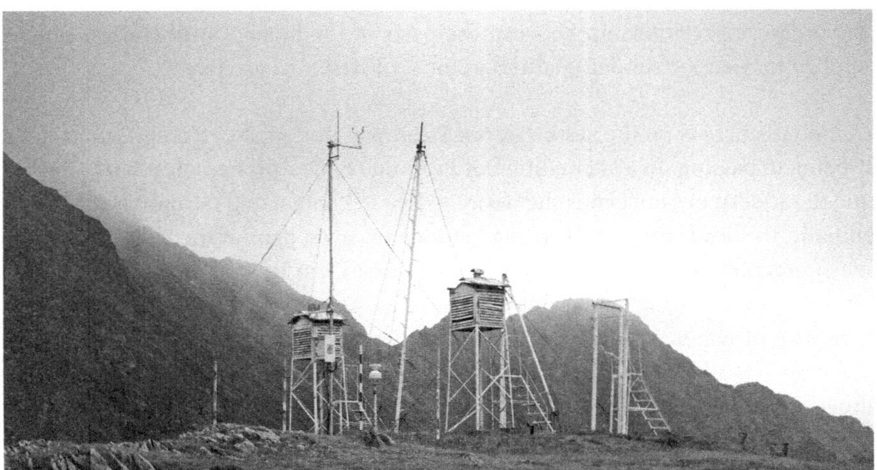

Figure 2.1 Technology revealing isolated nature. Weather station Bâlea Lake, Făgăraş Mountains, in central Romania.
Source: author.

> Moreover, there is increasing competition between places to present themselves as attractive to potential investors, employers, tourists and so on, to promote themselves, to sell themselves as service-, skill- and nature-rich places.[15]

So even an instrumentalized, economic materialism tends to bring the natural into awareness – and even to define and create the natural in any particular place and circumstance.

Design and nature: Two views

It is by now clear that nature signifies different entities in different contexts of discussion and action, and to different people. According to Macnaghten and Urry, nature must be negotiated. So, we are entitled to ask what nature is for – as a concept. I detect at least two main discursive uses of the term 'nature' in the context of architecture, landscape architecture and the built environment.

I have already alluded to the first use, which is *analogical*, drawing on parallels between biology and architecture in terms of shape, form and process. I also referred to the *evolutionary* metaphor, evident in the improvements of classes of artefacts over time. A second discourse about space and nature predates these, and it is a theme with which I began the previous chapter. It posits nature as a model or metaphor of balance, harmony and beauty to which we must be attuned. Steadman begins his account of biological analogy by explaining the classical, Aristotelian view of nature:

> The ideas of 'wholeness', 'coherence', 'correlation' and 'integration', used to express the organized relationship between the parts of the biological organism, can be applied to describe similar qualities in the well-designed artefact.[16]

The emphasis here is on the aesthetic, but Steadman does make reference to health and well-being in passing. In a later edition of *Evolution of Designs* he states that a 'consistent theme in modern organicism is the desire to live in contact and harmony with nature'.[17] Eventually, this leads towards 'a strong naturalistic, even pantheistic belief among some modern organicists in the symbolic role of natural form in architecture'.[18] As I discuss in Chapter 9, following the classical tradition, Alberto Perez-Gomez advocates an architecture of balance, 'allowing humans to live harmonious lives'.[19] By this reading, architecture should draw on nature to promote well-being. The aesthetic and the salutogenic converge.

The first engagement with nature outlined above, the analogic, now depends on algorithms, big data, and is at home with the idea of digital networks, mobile computing, social media and sensory feedback from the environment. With the second approach, it seems to me that salutogenic discourse encourages antagonism between the natural and the artificial. Salutogenic nature represents nature and our relationship to it as under threat, rather than nature to be copied, emulated, simulated or even enhanced. Smartphones, screen culture, databases, and algorithms, for all the benefits they might bring, exhibit the potential to upset our vital and health-giving relationship with the natural. For some, electronic systems and devices also deny the model of balance, equilibrium, health and well-being that nature offers. I mention these themes here as they pertain to discursive uses of the term *nature*: in this case to progress nature as a source of templates, patterns and processes to be emulated; second, as a source of health and balance, under threat in the digital age.

Biocentrism in architecture

Before progressing, it is worth reviewing the legacies connecting nature and architecture in my own discipline. Nature, the organic and the biological exhibit growth, decay and disorder, or at least an order that is outside of human mastery. Technology equates to function, control, precision, efficiency and factory production. It is easy to place modernity at odds with nature, and on the side of technology. But as any student of architecture knows, there is a thread of natural organicism in modern architecture. Louis Sullivan (1856–1924) provided exquisite organic detailing on building facades. At the same time we had various forms of Art Nouveau architecture. Frank Lloyd Wright's (1867–1959) organic domestic floor plans 'grew' out from the hearth, Bruno Taut's (1880–1938) fantastical expressionist buildings adopted suggestive biophilic forms. Buckminster Fuller (1895–1983) and Frei Otto (1925–2015) developed high-tech skeletal constructions clad in synthetic biomimetic membranes, to add to the canon of bio-architecture.[20]

As a demonstration of the contingencies on which we build our uses of nature, consider twentieth-century architectural politics. In the history of architectural modernism, adherents to biomimetic building were at odds with mainstream modernist sensibilities. According to Oliver Botar's comprehensive account of biocentrism in the arts, national socialism also embraced the organicist philosophy.[21] Botar attributes this support in part to the German agriculture minister at the time, and even to 'the vegetarian animal-rights supporter Adolf Hitler'.[22] In any case, biocentrism is associated strongly with romanticism, which in turn contains elements antithetical to many of the tenets of modernism, again, not least the political ideologies that romanticism reputedly fostered.[23]

But it seems that the seeds of modernism are permeated, impregnated, throughout with the biological, or at least, biocentrism. Botar asserts as much:

> The 'bio-centric' world view – with all its political complexities and contradictions usually swept under the rug – played an important role in the development of twentieth-century art.[24]

Leaders in the highly influential modernism of the Bauhaus, Walter Gropius, Mies van der Rohe and László Moholy-Nagy were complicit in modernist organicism, as were

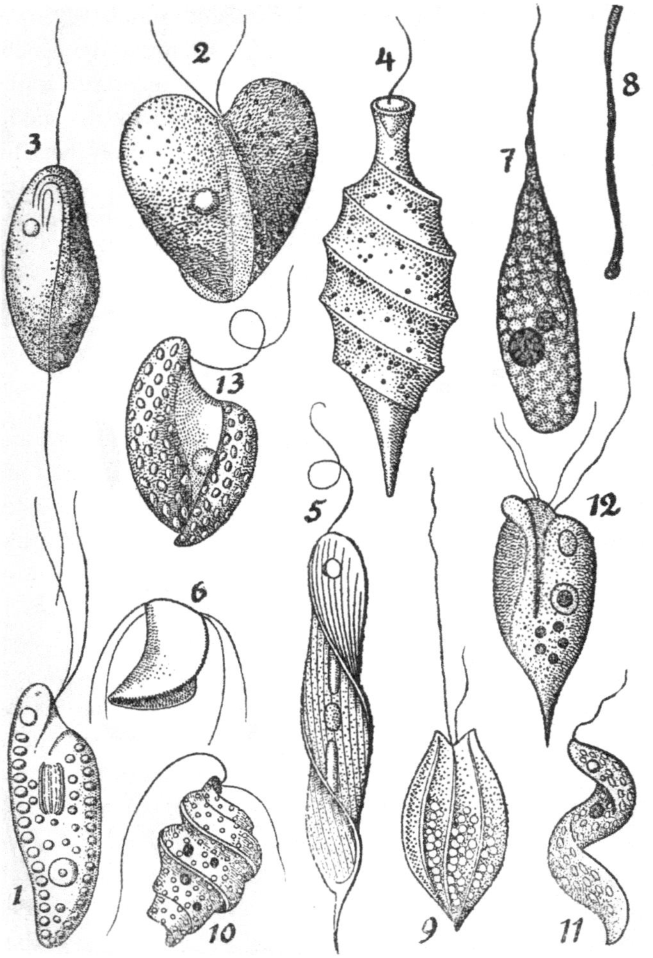

Figure 2.2 Illustration of Protozoa from Raoul Heinrich Francé's book *Die Pflanze als Erfinder* (1920).
Source: Wikimedia Commons.

Johannes Itten and Wassily Kandinsky among others. Many of the Bauhaus leaders apparently read the writings of the Austro-Hungarian botanist Raoul Francé (1874–1943). With seductive drawings of pods, cones and single-celled organisms that look like designed artefacts, Francé united the natural with the technological, via his assertion of the 'functionalist' character of nature: 'Nature's example is radically functionalist.'[25] According to Botar's account of Francé, 'humans had much to learn from organic technology and they stood to profit from its adaptation to their purposes, a process he termed Biotechnik.'[26] One of Francé's major works was called *The Plant as an Inventor*.[27]

Botar identifies several modernist threads that give witness to the influence of biocentrism and the biotechnical. They include the 'privileging of biology as the source for the paradigmatic metaphor of science, society, and aesthetics'.[28] Epistemology is based on a kind of psycho-biology. Biocentrism emphasizes nature, life and life processes rather than culture. It moves away from an anthropocentric world view and supports the agency and unity of all life, with an emphasis on flux, change and impermanence rather than stasis. It also emphasizes the importance of the whole rather than a reduction to parts. Much of this nature philosophy alights on the idea of oneness, a view that there is one substance, an interconnected unity, a philosophy manifested in its various flavours as *monist*.[29]

Blogger Joost Rekveld provides translations of fragments from Francé's German writing into English. Francé argued that the forces of nature tend towards a state of 'repose'. Form gets disrupted and reconfigured 'until the optimal, essential position of repose has again been reached and form and essence are again one.'[30] Biologist Johann Jakob Baron von Uexküll (1864–1944), who influenced semiotics, and Charles Sanders Peirce[31] also adhered to monism. So, there are connections between modernist biocentrism and biosemiotics. I will discuss biosemiotics in the next chapter.

Networks

So far, I have discussed semiotics as the communication of signs, but said little about networks. Networks are ubiquitous. 'Always think of the universe as one living organism', said Marcus Aurelius the stoic philosopher (and Roman Emperor); he also added, 'Remark the intricacy of the skein, the complexity of the web.'[32] Think of a network as a web. His sentiment is simple: if you only realized how interconnected your circumstances were to the rest of the world then you would be content with your small place in the organic order of things. There is a psychological, if not moral, dimension to net philosophy.

To identify a network, all you need to do is find some entities that you want to regard as distinct (for this purpose), and then establish some relationships between them. Geometrical and spatial relationships will do. So, there's a network in my neighbour's garden as the apple tree is next to the birdbath, which is next to the pond, and the apple tree is next to the plum tree, which is next to the rose bush. The network is more obvious if you draw it as nodes (the objects) and lines (arcs) connecting them (as the relationships). The net-scape gets more interesting if there are flows between the objects, or at least dependency connections. There's no requirement that the relationship arcs are of the same kind or scale.

It does not take much net thinking before you see networks everywhere, and of huge complexity, should we choose to identify and draw them, turn them into a circuit, or attach numbers to the nodes and arcs and perform calculations with them.[33] Enzymes, nutrients and electromechanical flows course through organisms. The web to which the stoics referred was probably less dynamic, contingent and open to invention than what I have just described, and more in line with the idea of perfect ratios, where elements are in ideal relationships to one another, such that the removal of one part upsets the symmetry of the whole.[34] Dependency networks in nature are sometimes like that.

Fertilizer runoff poisons ocean coral. The coral provides a habitat for other living things that are part of a food chain. Plastic waste turns into microplastic spores in the ocean that concentrate in fish, birds, marine mammals, and beyond. Effects propagate through networks. I do not think networks in nature are difficult to identify. Nor is it difficult to justify network models of relationships in nature, though they become complicated, depending on what the network model is for. Much has been written about complexity and chaos in natural systems.[35]

In the digital age, the things of nature are increasingly subject to digital flows.[36] A review article by computational biologists Saket Navlakha and Ziv Bar-Joseph indicates the similarities and differences between biological networks and 'the pervasiveness of mobile, wireless, and sensor devices'.[37] Sensor networks populate natural environments, as in cities. Nature is under surveillance with cameras secreted in trees, burrows and under the sea to monitor creature movements. Landscapes are mapped by remote sensing. Pets and wild animals are tagged. Databases and classification systems link together the things of nature. According to some, the network idea itself, as well as its exemplars, has undergone transition. Biologist and cultural theorist, Donna Haraway, identified a series of transformations from 'comfortable old hierarchical dominations to the scary new networks I have called the informatics of domination'.[38] One of the major differences now is that:

> No objects, spaces, or bodies are sacred in themselves; any component can be interfaced with any other if the proper standard, the proper code, can be constructed for processing signals in a common language.[39]

Networks are inevitable. Now they tend to unify, totalize and render things under a certain kind of control. There's a political agenda to Haraway's provocations, that feeds into her idea about the disruptive nature of the human–machine hybrid that is the cyborg, that apparently we are all in a sense becoming:

> I prefer a network ideological image, suggesting the profusion of spaces and identities and the permeability of boundaries in the personal body and in the body politic. 'Networking' is both a feminist practice and a multinational corporate strategy – weaving is for oppositional cyborgs.[40]

The network idea arguably draws from concepts of weaving, as expounded by Plato, and taken up by Deleuze and Guattari, among others.[41]

Navlakha and Bar-Joseph examined technical similarities and differences between digital and biological networks. Digital communications systems trade in long messages and 'high communication loads', best dealt with by algorithms that transmit data around the system at speed. Digital networks are overall fully connected and deterministic with identifiers that show where the message is coming from and the nodes to which it is directed. In contrast, messages sent through biological systems are much shorter than messages in digital networks, sometimes as simple as a binary on–off message to a neighbouring cell; they are robust but slow; incomplete and sparse with a random (stochastic) element.[42] Navlakha and Bar-Joseph illustrate communicative processes in the case of ants foraging for food. They explain in some detail how foraging ants communicate with each other as if passing packages of information through the internet.[43] Here, models of communication within biological systems draw on analogies with digital networks.

Digital networks have also been deployed to control biological systems. In the controversial art work 'Twitter Roach' by Brittany Ransom, multiple Twitter messages were processed and transmitted to a 'cockroach's body via a customized backpack wired into its small bodily frame'[44] to affect its movement, though human Tweeters were not entirely clear that they were influencing the insect's movements.[45] Such provocative network applications raise debates about human–animal relationships; these ideas will be developed further in Chapter 8.

Spatial semiotics

Network models are common in architecture, but what is the place of semiotics in design disciplines? Few would deny that architecture communicates, and in that sense, is a language, or at least like a language. As pointed out by the philosopher and semiotician Umberto Eco architecture does something else as well: it functions. A substantial tiled roof not only communicates protection from the elements, but functions to provide such protection. Occasionally the two become uncoupled: the roof can be made to look very substantial from the street, but away from view it is made of aluminium sheeting with a different functionality. The communicative aspects of the design differ from the functional aspects. The modernism of Walter Gropius, Mies Van der Rohe and others championed functional honesty.[46] But many such architects produced buildings designed to look as though composed of light steel elements that concealed substantial fire-proof concrete-clad frame elements set back from their facades. Buildings often do not function as they appear.

The 'misuse' of terms such as 'function' prompted architectural scholars in the 1970s to re-examine architecture as a language. According to a champion in this linguistic turn, Geoffrey Broadbent,

> The misuse, by architects and critics, of words such as *functionalism*, is itself sufficient reason for looking more closely at the language they use; at the relationships between buildings, the concepts which are used in discussing them, and the words by which those concepts are defined.[47]

He identified a further difficulty for scholars. They must select from a range of different schools of linguistic analysis that are incompatible with one another. One approach draws from the linguist Ferdinand de Saussure and his followers. Another focused on Charles Sanders Peirce, and a third would pursue Noam Chomsky's generative grammars. In fact, in his subsequent work, Broadbent veered to a critical engagement with structuralism, derived from Saussure, which led in turn to poststructuralism. Broadbent also co-wrote a book published in 1991 on deconstruction and the influence of Jacques Derrida.

What are the similarities between verbal and architectural languages? Building elements and their sub-elements and relationships are like words held together by grammatical rules to form phrases and clauses, which in turn go to make up larger linguistic units (sentences and paragraphs), and with sub-units within the words themselves (phonemes, morphemes). The composition of such elements into well-formed buildings according to rules hints at the idea of a grammar.[48] But Donald Preziosi identified a major difference in the 'signing medium'[49] between verbal (i.e. spoken or written) language and architectural language. In the case of verbal language, we are dealing with a medium that is 'relatively homogeneous'.[50] There are only so many letters and punctuation marks in English, for example, and a relatively limited range of basic elemental sounds. And after all, we are dealing just with marks on paper or sounds in the air. Architecture, on the other hand, along with the rest of the physical environment, deploys every material known to human kind in order to communicate. He said that 'architectonic signs are realized through what appears to be an impossibly complex hybrid of media',[51] involving 'anything drawn from the entire set of material resources potentially offered by the planetary biosphere, including our own and other bodies'.[52] Basic psychology in any case identifies verbal language ability with particular brain functions, if not regions of the brain. Engagement with the communicative aspects of space, place and architecture on the other hand would seem to involve a range of sensory modalities, perceptions, bodily actions and interactions.

The linguistic burden on architecture is of a different order then from that of verbal language. One tactic to address this challenge is to subsume verbal language within some larger field of semiotic competence, of which verbal language, architectural languages and anything else that we may wish to describe as a language, are a part. Structuralism adopts this wide view, and subsumes all of language, not just language as spoken and written. For the anthropologist Claude Lévi-Strauss, cultural linguistics involves taking whatever elements are to hand and putting them into certain relationships as part of a system of communication. Hunter-gatherer communities would use local animal species as a source of identification. So, for some parts of the community it would be bears and eagles. But a community in a different location might self-identify with lizards, swarms of wasps or the north wind. The relationships provide the medium of communication, rather than the elements themselves. According to structuralism, any objects displaying significant differences, in the appropriate context, will suffice as linguistic elements. As Preziosi said, such 'distinctions and disjunctions' can 'cue the perception of similarities and differences in meaning'.[53] On the other hand, the semiotics of Peirce treats sign systems as the overarching frame. Language is a subset of semiotics. As we will see, Peirce's model helps wrest nature from the restrictive entailments of the language metaphor, as if nature must always acquiesce to human cultural formations.

Semiotics sought to present as a science, which for many meant a system with agreed-upon categories (of relationships) and the ability to predict the characteristics of communities, their practices and architectures. Despite various attempts to do so, I think the field foundered in the attempt to identify linguistic units that were capable of being used instrumentally in understanding or designing buildings. As an architectural advocate of semiotics as science, Preziosi attempted such a schema, involving distinctive features, forms, templates, figures, cells, matrices and compounds, structures and settlements each organized within further categories and matched against verbal linguistic elements.[54] In the 1980s I, among others, hoped to apply such categories in computer-aided design systems.[55] The lack of instrumentality and the ambiguity and arbitrariness of semiotic categories easily gave way to digital data, and the more manipulable idea of parametric library elements.

Contrary to either of these positions, many architectural scholars, in the company of other cultural theorists, seized on the idea that language is a system of differences, articulated in the idea of the binary opposition. If bears and eagles can be deployed to indicate significant differences within hunter-gather communities, then so can concepts such as function and ornament, or function and communication, form and meaning, places and non-places, text and space, bike sheds and cathedrals, or any number of engaging oppositions, such as nature and artifice. Identifying such oppositions also raises questions about the relationships between such terms and whether and how one of them might be privileged above the other. The linguistic turn drifts inexorably into the ethical and the political, supported by Jacques Derrida's sophisticated arguments that constitute the philosophy and architecture of deconstruction.[56] Though not necessarily as amenable to instrumental manipulation as data-driven views about architecture, deconstruction and its variants have contributed to freeing up the design process, framing projects, enhancing critique, and not least providing vocabularies that keep people talking about architecture as rich, varied and potentially transgressive.

Other factors come into play in considering the drift from semiotics proper. Apart from the ascendancy of digital data, and in opposition to it, architecture has embraced phenomenology, which eschews the paraphernalia of instrumentalized linguistics or semiotics. While elevating language as 'the house of being', Martin Heidegger among others downplayed the primary status of the sign. Before there are linguistic units, that is, before there is anything to read in the environment, there is the experience of being in the world as an embodied being. Embodiment seems to win out in architecture over the esoteric entailments of semiotic discourse.[57]

As a thread in this book, I think of how semiotics rescues us from an obsession with digital data, at least in our relationship with nature, by showing that all things are caught up in networks of communication, even without human beings, other living organisms or technologies on the scene. Independently of his attempt to operationalize semiotics via categories and rules, Preziosi references basic biology and evolution in his account of the semiotic landscape. He constructs a helpful narrative around the idea that verbal language, gesturing and tool use developed together. Such speculations inevitably draw on the idea of human beings in direct contact with the natural world, and in which everything has the potential to 'be significant in some way'.[58] He refers to early humans 'as noisy, messy social mammals' who 'would inevitably leave semiotic droppings, metonymic marks and traces':[59]

> As highly social primates, early humans would be constantly assessing not only the finely-graded behaviours of each other, but also the continual flux of environmental information. In short, everything must be treated as potential evidence for something; everything must be addressed with a question; everything has a story to tell: but it is equally necessary to know what to ignore as noise.[60]

Preziosi here emphasizes the human's semiological relationship to environment, but later in the same book he says something similar about communication among non-human animals and environments.

For the architectural semiotician, buildings and building elements operate as signs, pointing to something other than themselves. So, for the semiotician one of the key roles of architecture is to *represent*. For scholars operating within the semiotic frame, the things of nature are among the targets of representation, evident in floral and foliated ornamentation, frescoes of nature scenes, shapes that resemble tree trunks, curves volutes, edges and patterns that somehow resemble or imitate (as icons) the things to which they refer. The architect and historian Charles Jencks was among those in the 1960s and 1970s who advocated for semiotics in architecture.[61] By the 1990s, such theorists had adopted the criticism and scepticism of the philosopher Jacques Derrida on matters to do with representation in architecture. Deconstruction and its challenges to supposed certainties, including the basics of a semiotic world view, were in full flow. But in the 1990s, Jencks affected a return to a newly reconstituted metaphysics, informed by the fields of quantum mechanics, and chaos and complexity theory.[62] According to Jencks, science and nature were leading the way, ahead of philosophers, semioticians and architects in bringing to light the nature of the universe we inhabit, and beyond. In this phase of semiotic exploration, Jencks posited a 'cosmogenic world view':

> It is the idea that the universe is a single, unfolding self-organising event, something more like an animal than machine, something radically interconnected and creative, an entity that jumps suddenly to higher levels of organization and delights us as it does so. Complexity Theory, the Gaia hypothesis, Chaos and Quantum theories all point in that direction, and they can give us great hope and strength.[63]

The idea that architecture *represents* something lingered in his narrative:

> The smooth growth of a wave form represents the continuity of nature, its unity and harmony, whereas the sudden twist represents the catastrophes of nature, the flip from one system to another, or the creative bifurcations which can bring progress as well as despair.[64]

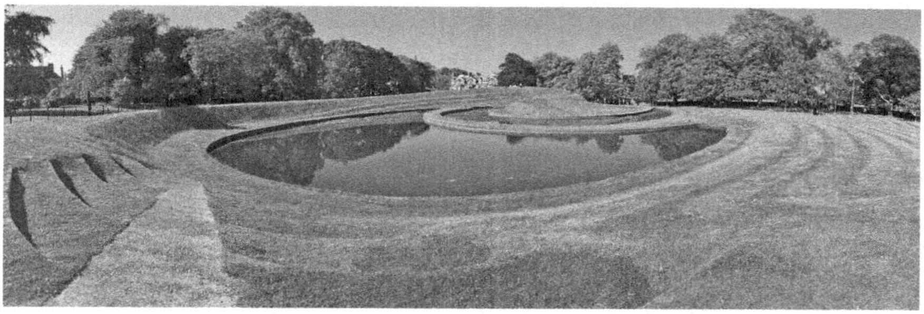

Figure 2.3 Charles Jencks's 'Landform'. Scottish National Gallery of Modern Art, Edinburgh. *Source*: author.

He adds, 'Since nature shows these two properties might not a cosmogenic architecture represent these two basic truths?'[65] By this reading, architects must take on the task of representing such up-to-date scientific understandings of the universe, a challenge that Jencks took on board in some of his own landscape garden design. That is a further discourse relating architecture and nature.

The issue of representation also surfaces in the discourses of sustainability, and of architecture in the age of the Anthropocene. Architecture could (and does in some cases) represent in some way the crises of climate change, pollution, deforestation, massive changes in landforms, and bring such matters to awareness in powerful ways through its forms, provocations and discourses. But architecture (among other complex, interrelated forces and specialisms) is clearly instrumental in shaping the world – not just representing it.[66]

I was drawn to the issue of representation in a chapter in the book *Architecture in the Anthropocene*, in which designer John Palmesino says, 'It doesn't work, that's the problem, the entire take on architecture as representation; as opposed to interference, constructive practice, and making things up.'[67] The idea of the Anthropocene had not yet impacted on architects in the 1990s when Jencks wrote *The Architecture of the Jumping Universe*. The Anthropocene brings to the fore the idea that architecture changes nature rather than simply representing it, and unleashes powerful and sobering modifications to the discourses relating architecture to nature, to be discussed further in Chapter 6.

Nature and pragmatism

I have already alluded to a pragmatic position on nature and its meanings. In my book *Designing Information Technology in the Postmodern Age* I demonstrated the impact of pragmatism on the world of digital media.[68] Others have developed the theme further in relation to 'user experience design'.[69]

Peirce started the philosophical movement known as pragmatism. He defined himself as a logician, and populated much of his writing with logical proof.[70] He introduced the simple proposition of pragmatism via a difficult maxim:

> Consider what effects that might conceivably have practical bearings we conceive the object of our conception to have: then, our conception of those effects is the whole of our conception of the object.[71]

If the sentence seems obtuse, then focus on 'practical bearings'. An example may also aid comprehension. Think of a concept: justice, god, nature, technology, architecture or a table. The pragmatic maxim asserts that a concept has no other scope or meaning than the practical difference it makes to hold to that concept. You do not have even to believe in the concept for it to have a practical bearing. The concept of nature provides a useful example. Whether I believe in it or not is less relevant to the concept than that others and I find practical use in the concept. As far as the concept of nature has effects such as to transport people into the countryside, stay indoors, establish national parks, adjust agricultural practices, stimulate healthy lifestyles, sell products or provoke discourses about technology, these effects constitute the concept of nature. There is no concept of nature beyond the uses to which we put the term. Pragmatism takes exception therefore to the idealistic notion that concepts somehow exist independently of our acting with them. This point carries certain controversies: not least, the pragmatic orientation seems not to allow space for the things of nature to simply be. I will elaborate on this charge in Chapter 9.

Pragmatism resonates with concepts within phenomenology, and Martin Heidegger drew extensively on everyday practical engagement with equipment and practices as the basis of his philosophy of being-in-the-world.[72] According to one of his commentators, Hubert Dreyfus, Heidegger was familiar with American pragmatism,[73] though he does not refer to Peirce in his writings. As I will indicate in subsequent chapters, there are differences between Peirce's pragmatism and phenomenology.

Architectural and other disciplines connected with place also draw on pragmatism. As I have rehearsed in other publications,[74] the test of any set of ideas is what difference those ideas make to the way architecture, or any discipline, is practised, talked about, assessed and taught. In their professional lives architects and environmental specialists need to be concerned less with the question, 'is it true?' than 'what practical difference does it make?' This mode of interrogation is practical, examining how the work of the philosopher fits within a context of the practices of a discipline: designing, documenting, building, reflecting, evaluating, interpreting, critiquing and defending, as well as formulating histories and learning about architecture, landscape and the nature of place.

This excursion into the intellectual context of our discussion has highlighted the challenges evident in defining nature, architecture's biocentrism, the nature of networks, semiotics and space, and pragmatism. I turn next to data and elaborate on its conflict with a semiological view of communication in nature.

CHAPTER 3
THE BOOK OF NATURE

Walk through banks of wildflowers, feel the chill of the setting desert sun, marvel at the horizon from the seashore, and rub your flesh against the textures of wild landscapes. People experience nature under many conditions. Of course, it is never just raw nature. Stories, pictures, documentaries and other cultural artefacts prime our expectations and influence what it is we see, hear, feel and smell. Tools, devices and technologies also extend this experience of the natural. Digital tools are prominent among the technologies that bring aspects of nature into human experience. They reveal, conceal, distort, magnify and sometimes make the experience of nature.

The Hungarian-born US semiotician and linguist Thomas Sebeok (1920–2001) reminds us that rather than raw 'reality' we are dealing with 'nature as unveiled by our method of questioning. It is the interplay between "the book of nature" and its human decipherer that is at issue.'[1] In this chapter, I will begin with data; then return to semiotics. As I will show, big data presents an image of nature as distinct, distant and under human control. The more nuanced semiotic orientation presents an integrated understanding of nature in keeping with the desire of John Dewey and other pragmatists to thicken *experience* and show how it 'penetrates into nature and expands without limit through it'.[2]

Big data

Researchers into the things of nature often agree about the important role of mediating technologies. Geographers Harvey Miller and Michael Goodchild assert:

> Instead of looking through telescopes and microscopes, researchers are increasingly interrogating the world through large-scale, complex instruments and systems that relay observations to large databases to be processed and stored as information and knowledge in computers.[3]

To be useful in computer calculations, data presents as tabulated information, such as an inventory of employees in a firm, with fields indicating address, salary scale, years of service. It can also be 'relational', where a database about different cities and their characteristics is linked to information about individual inhabitants. A list of facts or simple predicates can also be data: 4,827 people live in this town; the average global temperature is 14 degrees centigrade. Data is never just evidence in the raw. It must be collected selectively, sifted, processed for some purpose, and is frequently bound up in a theory or a frame of reference.

Miller and Goodchild note that current data flows from all the monitoring devices, transaction records and surveys coursing through our digital systems amounts to more data than we can analyse. I take this to mean there is even more data than any individual or team with a spread sheet can use without the aid of sophisticated statistical analysis, smart algorithms, high bandwidth and powerful processing.

Data can serve as powerful evidence and can clarify matters, but it also poses challenges. Anyone with a personal computer knows about the problems with data, not least its tendency to grow at an unnerving rate, take up space, clog bandwidth as it is moved around and demand upgrades of the hardware and software that support it. Emails and tweets are data. So are picture files, sound tracks and movies. Sometimes we store data, or it flows through our world as if a stream. Data is on the move. We pull and scrape data from the web. It gets pushed at our appliances, smartphones, tablets, television sets, mobile computers and workstations. We generate it as well, even without knowing. Then there is all that data of which we are scarcely aware but that affects us anyway – financial, travel, meteorological, surveillance, medical, remote sensing, monitoring, security and scientific data. The big-data exemplars are familiar to most of us observers of digital and social media – the flow of Twitter feeds, Facebook news feeds, bus and train travel information, airline schedules, and real-time weather information. But that is just the tip of a massive iceberg, or glacial tributary, or spray from a deluge of data gathering from the natural environment, oceans, cities, people and outer space. Dealing with the scale of it, its demands on storage and bandwidth, its redundancy, its inconsistency and its heterogeneity (data is in many different formats) constitutes major technological challenges. But since about 2008, the data maelstrom has changed its hue. That change in perception about data is the so-called *big-data revolution*.[4]

The problem of storage is addressed in part by the idea of 'the cloud'. The data is stored, backed up and ferried between very large file servers, and data users do not need to know how or where that happens. But the scale, flow and heterogeneity of all that data constitutes a challenge, the mastery of which informs commerce, high-level decision-making, and politics.[5] In terms of the physical environment, urban and environmental control systems, the sciences, policy formation, and aspects of design have purchase in the big-data revolution, not least via the vast streams of pictures, sounds and documents that hone the performance of automated image, voice and text recognition algorithms.[6] Such data also has uses beyond its immediate purpose. All those photo sharing and retail transactions do not only move pictures, goods and money around, but can be mined for information about consumer habits in general. Medical records contain information about individuals that can be correlated with data from other sources to yield insights on the state of people's health, and the distribution of diseases over time and space. A 2008 US report by Tony Hey and colleagues called *The Fourth Paradigm: Data-Intensive Scientific Discovery* makes the case for improving digital infrastructures to support the flow, storage and analysis of massive quantities of scientific data.[7] Several articles in that volume reference the provocative and influential article in *Wired Magazine* by the then

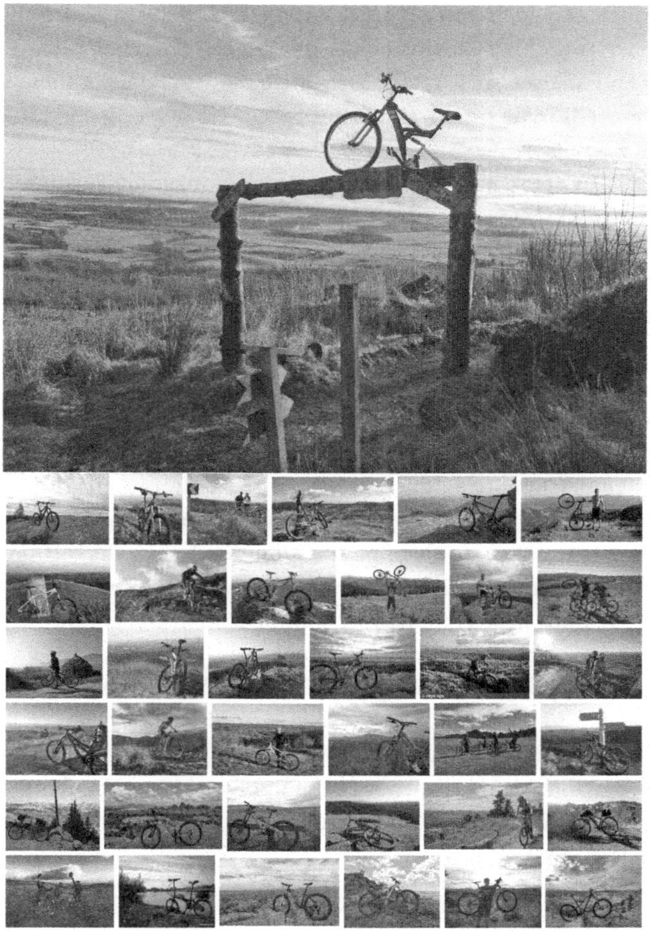

Figure 3.1 Search results from Google Images. The top image (Crieff, Scotland) is by the author. Those beneath are matches found by Google Images search.
Source: author.

chief editor Chris Anderson.[8] Anderson's article is entitled 'The end of theory: The data deluge makes the scientific method obsolete'. Anderson argues that big data ushers in a step change in thinking about how knowledge growth and science happen.

I think these big-data issues are brought into sharp relief as we think of natural, geographic, urban and other spatial data.[9] Consider a digital map charting a couple of hectares of the countryside at varying levels of detail perhaps down to 1-centimetre resolution. There will be height data, a record of material properties at different depths, including archaeological data, beneath and above the surface, and recorded over past, present and projected periods. People, animals, machines and vehicles move across, over and through these surfaces. Now expand that data pool to a whole continent that includes cities, industrial complexes, transportation systems, natural resources, the weather and ecosystems, much of which gets recorded via Light Detection and Ranging (LiDAR) laser scans generating three-dimensional point clouds of spatial data. It is possible to capture, store and analyse such spatial and attribute data. Data is there in volume. To be of any use it needs to be stored, sifted and processed at speed, and it is likely to be in many different formats.

But big data is an ocean not only of data but also of magnified claims. As an example of big-data benefits Victor Mayer-Schonberger and Kenneth Cukier indicate how 'improving and lowering the cost of healthcare, especially for the world's poor, will be in large part about automating tasks that currently seem to need human judgment but could be done by computer'.[10] Big data will be able to tell us when we are about to fall ill. They add, 'Soon big data may be able to tell whether we're falling in love.'[11] Their book follows a popular-techscience format, beginning with sensational claims about the revolutionary significance and promise of the new technology. They follow this up with warnings about misuse. Then they exhort readers and politicians to increase their knowledge and understanding of the technology. The authors then advocate for cautious and judicious use of the technology. Finally, they advocate for further development, which is to say more money needs to be spent on developing the technology. They say that big data requires 'new principles by which we govern ourselves'.[12] Privacy surfaces as a major issue: 'We must protect privacy by shifting responsibility away from individuals and toward the data users – that is, to accountable use.'[13] They advocate a sense of human volition and responsibility in decision-making, and caution that we need to provide analytical expertise to assess big data. Otherwise, there is a danger that the world turns into a 'black box, simply replacing one form of the unknowable with another'.[14] They acknowledge that big data introduces a raft of problems, though they neglect one important solution, which is to subdue and moderate the claims in the first place, so that less is promised or expected.

Data bias

For all its benefits, big data does not serve everyone equally. Like digital bandwidth, access to the web, healthcare and many other social goods there are the beneficiaries and those who are at the margins. Any innovation or resource promoted as a good has the potential to amplify the difference between those individuals and nations who can afford it and those who cannot. Many scholars challenge the utopian speculation that universal global benefit follows the big-data tidal wave. Critics highlight data challenges, of which I suspect most people working at the coal-face mining big data are aware. The first and most obvious is data bias. Any student of social science knows that the collection of data favours certain interests and diminishes others: for example, the collection of data on passenger usage on bus routes suggests that routes used less frequently will be discontinued. The collection of the data skews decision-making in a certain direction, in this case in favour of a story about the many (very easy to count) as opposed to those with the greater need (very difficult to measure). It is hard to argue against the statistics that data provides.

As a second criticism, data gathering skews behaviour. Where people are involved, data collection can also direct behaviour. Consider university space audits where pollsters go from room to room counting attendances at lectures. Lecturers encourage students to attend or they will lose the space. As evidenced by inaccuracies in political polling, knowing that the data is being collected influences people's behaviour.[15] The third criticism orbits around the issue of privacy. Where people are involved there are abundant privacy issues. If I were asked in a market survey how much I spend per month on haircuts, tins of tomatoes, petrol, alcohol and online services then would I even know, and would I be prepared to disclose this? Most consumers, travellers and social media users would be guarded about volunteering the kind of information that gets collected about us automatically at the supermarket, at border controls or on social media.

As a fourth category of critique, big data is supported by the profit motive. Much of the data in big data is collected and managed by companies with a stake in the data and its uses. Edd Dumbill recognizes this in his warnings: 'It's no coincidence that the lion's share of ideas and tools underpinning big data have emerged from Google, Yahoo, Amazon and Facebook.'[16] These are commercial organizations after all. Companies want to maximize revenue, much of it through promotion and advertising. Big data follows big markets. It also fuels the idea of data-rich city planning and management, and the 'smart city.'[17] In an article critical of smart city projects that exploit big city data, Rob Kitchin identifies a 'neoliberal ethos that prioritises market-led and technological solutions to city governance and development'.[18] He asserts, 'It is perhaps no surprise that some of the strongest advocates for smart city development are big business (e.g. IBM, CISCO, Microsoft, Intel, Siemens, Oracle, SAP).'[19] Big data rehearses again the problems of global capital, commercial products dressed up as social goods, and the marginalization of those unable to participate in the supposed benefits.

As a fifth criticism, data does lure us in particular directions. Many designers, artists and others in the arts and humanities may be sceptical of big data, but such fields do not carry the same influence in political decision-making as the worlds of science, management, finance, and governance. In these areas, data is highly influential. Already I see research drifting towards projects that have some traction in the world of data. Arguably, big data and its emphasis provide further bias away from the arts and humanities and towards disciplines more comfortable with whatever can be counted and quantified. The supposed digital humanities provide another indication of this drift from the tenets of the interpretive arts, or at least the human art of interpretation, and other forms of evidence-based research.

Big data has also operationalized aspects of the humanities. It is a commonplace to note that texts exist in vast quantities. Publishers in the UK alone produce over 180,000 book titles each year, with about one-third in digital formats, and there are all the other words generated online – self-published, or unpublished – and journal, magazine and newspaper articles. These large text corpuses can be treated as big data – counted, mined, probed, analysed, compared, correlated and turned into tables, graphs and network diagrams, even without the need for anyone to interpret or understand any of it. More precisely, scholars can use computer programs to transform literary content into different formats to understand it better – or at least differently. That is the so-called 'distant reading', as opposed to close reading.[20] The scholar stands back as if from afar and reviews a whole corpus (collection) of works, and combinations of corpuses. It is less about singular texts, and more about whole collections (e.g. the complete works of William Shakespeare, all nineteenth-century English novels, or the Hansard Reports). Franco Moretti of the Stanford Literary Lab hopes to find the 'unified theory of plot and style',[21] as if gathering data from the natural world. Kathryn Schulz in the *New York Times* is suitably sceptical about this kind of study.[22] She makes the point that literary data is created 'by design', and not subject to the independent, distant readings science claims to make of natural phenomena. Dispassionate analysis of texts can only get us so far before we have to commit to the meaning of what it is we are reading or do not have time to read.

According to ethnographer Tricia Wang, in light of big data, the tenets of qualitative and ethnographic research are also under threat in various quarters. She argues in a blog posting: 'Our work can be all too easily shoved into another department, minimized as a small line item on a budget, and relegated to the small data corner.'[23] In her fight-back post she asserts that 'Big Data Needs Thick Data', a rallying cry to understand big data-its narratives, benefits and pitfalls. The tendency towards quantification and other big-data excesses are practical limits to data, of which I think most people working in fields that rely on data are aware. The issues I have outlined so far are not alien nor are they particularly controversial to anyone working with data.

But the sixth category of criticism revolves around the claim placed at the door of big data that big data provides access to theories; more specifically it dispenses with the need for theory as conventionally understood. Big-data enthusiasts construct a philosophy around big data. The strong claim is that big data constitutes a repository developed without bias or favour, to be deployed for whatever purpose we choose. The data is there, streaming around us as if automatically to be captured and accessed, even if we cannot yet think of a use for it, or what it might show. The mantra is that data is neutral. If there is enough of it then it covers every individual case and accommodates just about every way of looking at the phenomena it represents. Statisticians do not need to sample. They have the complete set. It can be probed to identify the most typical, marginal and exceptional, and even provide generalizations, theories and models, though these are no longer necessary.

Anderson's article on big data, 'The end of theory' promotes the putatively diminished status of theories.[24] Theories are an efficient substitute for stored data, but they leave out a lot – all those exceptional cases. He argues that now we can retain the raw data to be interrogated over again. We see this in forensic investigations where physical and electronic evidence is retained for several decades. The case may be reopened as new techniques for analysing the data emerge. We have seen as much in the case of DNA samples where a conviction follows many years after the crime event.

The case against theory argues that big data does not presume theories and the biases they entail. Data ought to be freely available (anonymized) and available to different communities of researchers. So, there is a call for open access to data. Enthusiasts think that big data is democratic. It also lends itself to crowd sourcing, the recruitment of armies of volunteers to generate, interpret and deploy data, and give access to everyone. As a case in point, Gillam et al. refer to what they call the 'health care singularity'.[25] This is the moment when medical knowledge becomes 'liquid'.[26] Vast numbers of patient records will aggregate in patient data clouds to be tapped, and from which hypotheses can be extracted and predictions made. This ascendancy of data will also change academic research practices. Gillam et al. assert: 'To enable instantaneous translation, journal articles will consist of not only words, but also bits. Text will commingle with code, and articles will be considered complete only if they include algorithms.'[27] Data is of little use without algorithms to process it.

In this and other respects, and for all its newness, it seems some people position big data within an old-fashioned philosophy. Enthusiasts such as Anderson presume that the data is as useful, complete and good as the object being studied. It is as rich as the physical and social world around us, or at least it can be treated in the same way. For those who follow this belief in data it makes sense for Dumbill to assert that there is an immediacy to big data that removes the need for theories: 'Having more data beats out having better models: simple bits of math can be unreasonably effective given large amounts of data.'[28] He is mainly thinking of retail, 'If you could run that forecast taking into account 300 factors rather than 6, could you predict demand better?'[29] But if you translate such claims to the world of nature then we have a re-enactment of the platonic idea that everything is underpinned ultimately by number, mathematics, code and logical predicates,[30] and that the world is data after all.

Nature is not data

Big-data discourses give no priority to nature. The means of gathering data vary, but measuring and recording the weather, movements of the earth's crust, the migration of birds, and other natural phenomena are no different from monitoring traffic flows in cities, the distribution and fluctuations of property prices, and the behaviour of shoppers. Big data expands the observation advanced by Martin Heidegger that for all the benefits afforded by advanced technologies everything is treated as the same, ultimately a resource for human exploitation, or at least its potential.[31]

Nature and data do connect through the idea of code, not least bolstered by advances in understandings of DNA. But some of the problems and conceits I addressed in the previous section stem from the pursuit of big-data substrates and one-to-one data maps of the universe that can be run through algorithms in the computers of the future, offering access comparable to nature in the raw.

Add to the critique of this nexus between data and biology, the weight of objection to both the informational model of communication, and its extension to DNA sequencing, DNA as 'nature's code'. Lily Kay's book, *Who Wrote the Book of Life? A History of the Genetic Code*,[32] presents a case against a literal understanding of nature and its codes.

Not least, there have been several important political motives in promoting models of biology based on information theory, cybernetics and systems theory as biotech industries ride on the coat tails of the undisputed ascendancy of computer networks in contests between nations following the Second World War. I will say more about DNA and data in Chapter 4 on 'biohacking'.

Biosemiotics

Semiotic discourse begins with something more practical and grounded in human experience than raw data. As I have outlined already, the pragmatic philosopher Charles Sanders Peirce advanced semiotics as a field of study.[33] William James and John Dewey frequently quoted Peirce. Thomas Sebeok drew on Peirce's writings to provide a more recent and authoritative account of the field of semiotics and its application to the world of nature.[34] For Sebeok, the theory of semiotics begins with living things, but extends beyond that. The minimal semiotic unit is the living cell found in microorganisms, colonies, plants and animals, including human beings. Human bodies are 'assemblages of cells, about one hundred thousand billion (10^{14}) of them, harmoniously attuned to one another by an incessant flux of vital messages'.[35] Messaging is key in organic systems, and there are many means:

> The genetic code governs the exchange of messages on the cellular level; hormones and neurotransmitters mediate among organs and between one another (the immune defence system and the central nervous system are intimately interwreathed by a dense flow of two-way message traffic); and a variety of non-verbal and verbal messages conjoin organisms into a network of relations with each other as well as with the rest of their environment.[36]

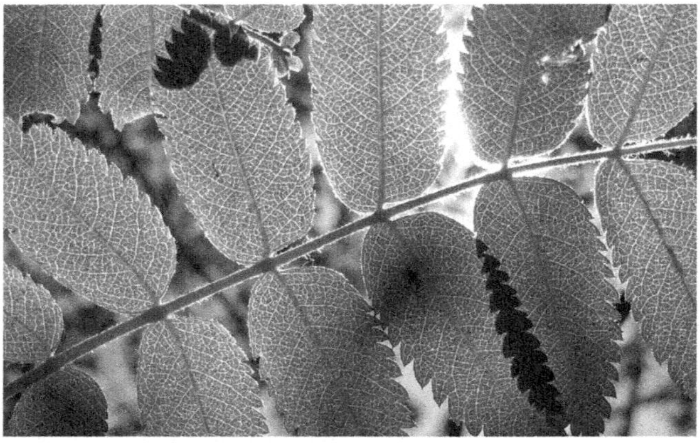

Figure 3.2 The vein structure of a pinnated leaf visible in sunlight to the naked eye.
Source: author.

The term 'data' occurs in Sebeok's key text on semiotics only five times, referring to medical diagnosis, in favour of the operative term 'message'. He makes no overt case against data, but reading between the lines the primacy of messaging leaves open the possibility of complexity and contingency in communication, of which the idea of data is a rarefied abstraction, suited to particular purposes rather than an ineluctable substrate to the world of nature. After all, anyone having worked with data, tabulated and coded it, subjected it to statistical analysis and translated and charted it, knows that data mining is a contingently human practice. The process needs to be tailored to the ends in view.

Reading nature

The theory of signs (semiotics) is interesting not least as it repositions the discussion of nature away from the reductive notion of data towards experience, and arguably the totality of experience.[37] Whereas notions of data draw on machine metaphors, semiotics takes its cue from the world of nature. Even contemporary urbanized human beings know about reading nature's signs: 'Red sky at night, shepherd's delight. Red sky in the morning, shepherd's warning.'

Building on Peirce's work, and that of the biologist Johann Jakob Baron von Uexküll (1864–1944), Sebeok identifies six types of signs. Each sign type references nature in some way: the signal, symptom, icon, index, symbol and name. Sebeok recognizes the overlap between sign categories, and the circumstances in which they operate, which will differ depending on circumstances, practices and the interpretive frame in which the sign user/receiver operates. The categories in any case help structure debates among semioticians, and have interesting implications for the way we 'read' nature.

The first sign category is the *signal*. Signals trigger some action in a receiver. Peacocks signal their impending mating attentions to a peahen via visual display, gestures and sounds. The first frost signals to a bear to prepare for hibernation. For Sebeok:

> Signaling activity, in its simplest form, is produced by an individual organism; it represents information; it is mediated by a physical carrier, and it is perceived and responded to by one or more individuals.'[38]

Not all signals are backed up by intentions, or provide advantage to the signaller. In the animal world of zoo-semiotics inadvertent signals are passed across species and between plants, animals and other elements of the natural world. After all, the physical environment is replete with signals. For us humans it is the colour of the sky, the length of the days, the temperature, sunlight, birdcalls, the sound of traffic, the chime of a clock, traffic lights, a pop-up message on a smartphone or a signal from the supervisor that it is time to leave the factory.

The second sign category is the *symptom*. This is the familiar involuntary sign that there is a change in the state of an organism: a grey pallor (ill health) or a rosy complexion (good health). In contemporary clinical practice a symptom is usually self-

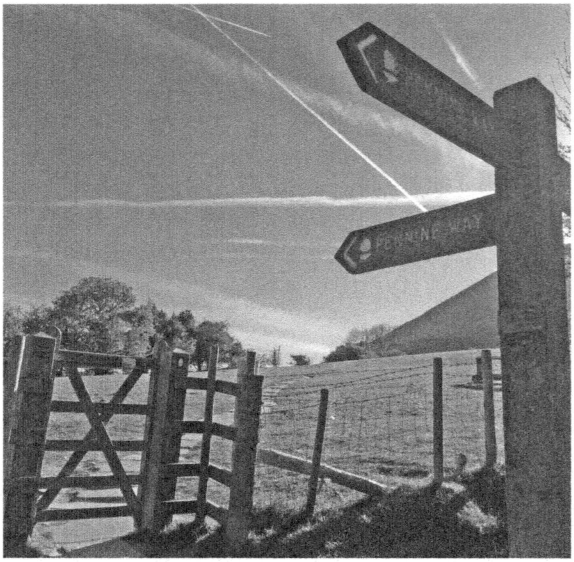

Figure 3.3 Overlay of signs in the countryside. Pennine Way, Derbyshire, England.
Source: author.

reported; a sign is something observed. But in general terms a symptom is evidence of something that is otherwise hidden or at least not obvious. In a cultural context we could say that online abuse on social media is a symptom of deeper social problems such as poor educational provision or lack of a stable family environment, or the obsession with big data is a symptom of the human need to feel in control. The use of the general word 'sign' seems to have its origins in the idea of symptom in early medical practice, that is, a sign is related to illness, dating back to Hippocrates (460–370 BC) according to Sebeok.[39] In their seminal book *The Meaning of Meaning* published in 1923, Ogden and Richards reinforced this clinical connection via the example of diagnosis, for example, high body temperature as a sign of influenza.[40] The importance of the symptom in the history of sign theory is further indication, if we need it, of the relationship between signs and the things of nature – in the most obvious case, the health status of an organism.

The third sign category is the *icon*. This is where a sign appears as a drawing or image that bears some similarity to the object it designates. According to Sebeok:

> An icon is a sign that is made to resemble, simulate, or reproduce its referent in some way. Photographs may be iconic signs because they can be seen to reproduce their referents in a visual way.[41]

He adds that onomatopoeic words, such as woof, meow, tweet, belch, buzz and boom, are also iconic signs 'because they simulate their referents in an acoustic way'.[42] He also refers to: 'Commercially produced perfumes that suggest natural scents as iconic.'[43] Sebeok provides examples from the animal world. Ants emit an alarm signal via a chemical, the intensity of which mirrors the intensity of the threat: 'The sign is iconic inasmuch as it varies in analogous proportion to the waxing or waning of the danger stimuli.'[44] Then there are defenceless insects that mimic other more dangerous species (e.g. wasps) in order to deceive the birds that would otherwise eat them. In an icon there is an attempt to mimic the thing being referenced.[45] Peirce examines images, diagrams and metaphors as further subspecies of icons, though Sebeok thinks this further categorization is less necessary than the categories he identifies.

The fourth sign category is the *index*. An indexical sign bears a direct relationship with the thing it references. Sebeok refers to Peirce: 'An index, as Peirce spelled out further, "is a sign which refers to the Object it denotes by virtue of being really affected by that Object."'[46] Forensic investigations in crime stories inevitably appeal to the idea of the indexical sign: the bullet hole in the wall is a sign that someone fired a gun; the horseshoe found on the road is a sign of a well-bred young pony (according to an inference by Sherlock Holmes). Indexical signs are therefore matters of 'fact' once established. According to Sebeok: 'Temporal succession, relations of a cause to its effect or of an effect to its cause, or else some space/time vinculum between an index and its dynamic object, ... lurk at the heart of indexicality.'[47] As the index operates via a direct relationship, nature is replete with indexical signs:

> Signs, inclusive of indexes, occur at their most primitive on the single-cell level, as physical or chemical entities, external or internal with respect to the embedding organism as a reference frame, which they may 'point' to, read, or microsemiotically parse – in brief, can issue functional instructions for in the manner of an index.[48]

In everyday language we think of how A causes B, though causality is just one manifestation of the indexical sign. Sebeok thinks that the function of the indexical sign is *superordinate* to causality, that is, it is more important, a higher category of phenomena.

The index comes before any attribution of meaning. According to philosopher Albert Atkin in his review of the indexical sign, 'an index offers no description of its object'.[49] As I indicated at the start of this chapter, Sebeok invokes the idea of 'the book of nature' as dealing in indexical signs,[50] for example, hunting practices (following animal trails) and divination (reading animal entrails or the flight of birds to tell the future),[51] as well as the idea that nature is written in a kind of code requiring decipherment.[52]

The fifth sign category is the *symbol*. The symbol is well known as a communicative medium among humans: brands, insignias and crests that indicate something beyond their mere representation. We might think that only human beings resort to symbols, but Sebeok asserts that 'the capacity of organisms to form intentional class concepts obtains far down in phylogenesis'.[53] He talks about actions among animals that are otherwise unconnected with any function, such as when a dog wags its tail. The same gesture in the world of cats means something different. He identifies a species of fly (*Empididae*) that forms secretions into a balloon shape that it then offers to a potential mate. Some gestures and objects have sign functions that cannot come into account as signals, symptoms, icons or indices, and are ostensibly useless, except for what they mean by the actions that they invoke.

The sixth and final sign category is the *name*. We humans may give ourselves not only fixed names, but also multiple identifiers, nicknames, IDs, handles, titles and labels. Naming is ubiquitous in the human world, but for Sebeok,

> it is well known that all animals broadcast a steady stream of 'indentifiers', that is, displays identifying their source in one or more ways: as to species, reproductive status, location in space or time, rank in a social hierarchy, momentary mood, and the like.[54]

Human and non-human animals identify each other via naming. In the case of non-humans it is through different senses and time frames.

So, we read the book of nature, and it reads itself, in these six ways at least, as the signal, symptom, icon, index, symbol and name, and there are other subcategories we could identify beneath these. Semiotics draws substantially on the idea of classification (families, genera, species and subspecies of signs), and at times looks similar to the task in which we identify and classify biological or other natural specimens. In fact, the term 'superordinate' introduced above indicates a concern with structure, how the name of the genus comes before the name of the species in a hierarchical system of derivation and dependency. In these and other ways, semiotics connects with the academic study of nature as an exercise in classification.

From a hermeneutical perspective, any such sign system is subservient to the workings of interpretation. Though intent on a science of signs,[55] Sebeok highlights how one form gets translated into another depending on circumstances. These sign categories are aspects of communication rather than distinct modes. So a red sky at night can act as a signal to shepherds to let the sheep graze further afield, it is a symptom of high pressure in the upper atmosphere trapping dust particles, it is an icon in that the more intense the barometric pressure the more blue light gets filtered out (leaving red) that is, the redness is an analogue of the barometric pressure, it is an index in that the redness of the sky has a cause, though whether or not the redness is a good predictor of clement weather would depend on further analysis. The red sky serves as a symbol of rest, repose and new life, and it serves as an identifier (name) for the start of the night. Such is the conditional semiotic structure of a red sky at night.

Cognition in the wild

By way of summary, the 'science of signs', semiotics, provides valuable insights into the relationship between the world of nature and the highly artificial world of networked computers. Permit me to extrapolate from Sebeok and other theorists of semiotics and apply their insights to data. First, we must abandon the idea that at the core of artificial, animal and plant communication systems we have data – at least, that is the proposition I am pursuing in this book. Contrary to the semiological view, and according to a common computational typology there is data (just 1s and 0s, or perhaps unstructured facts), from which we might infer information when put into a meaningful context. Further up the scale there is knowledge. Some even add wisdom above that. This is the so-called 'DIKW model' (standing for data, information, knowledge and wisdom) popular in information science.[56] Included in its many practical shortcomings, the DIKW model drives life out of nature. It also shows that contrary to expectations computer systems that attempt artificial life (AL),[57] and supposedly intelligent robots,[58] are maddeningly un-alive. Think instead of signs as the unit of communication, with data as a derived and manageable unit of calculation suitable for computer processing.

To further the case against data as a defining factor in nature I include among these semiotic texts the fascinating book *How Forests Think: Toward an Anthropology Beyond the Human* by the anthropologist Eduardo Kohn.[59] Kohn does not write about data, but it is clear that the concept of the sign delivers something that data cannot, leading him to assert that 'life is inherently semiotic',[60] and 'the logic of evolutionary adaptation is a semiotic one'.[61] Drawing on Peirce's work, evolutionary theory and careful observations of life in the forest region of Ávila in Ecuador, Kohn develops the proposition that there is a unity among living things that otherwise eludes definition, that affirms the capacity of thought beyond the human (albeit on different time scales) and that restores some of the enchantment of nature. It is worth quoting from the book, to capture the tenor of his propositions about life, with his deliberate hyphenation of 'mean-ings':

> If thoughts are alive and if that which lives thinks, then perhaps the living world is enchanted. What I mean is that the world beyond the human is not a meaningless one made meaningful by humans. Rather, mean-ings – means-ends relations, strivings, purposes, telos, intentions, functions and significance – emerge in a world of living thoughts beyond the human in ways that are not fully exhausted by our all-too-human attempts to define and control these. More precisely, the forests around Ávila are animate. That is, these forests house other emergent loci of mean-ings, ones that do not necessarily revolve around, or originate from, humans. This is what I'm getting at when I say that forests think. It is to an examination of such thoughts that this anthropology beyond the human now turns.

What becomes of thought when we admit it as something that forests do? That forests might think, speak, whisper and have agency survives in remnant form through many fantasy narratives, not least the story of *Babes in the Wood*,[62] and J. R. R. Tolkien's forest of Fangorn in *Lord of the Rings*.[63] In the latter the trees speak to one another in words. As a more serious proposition, the communicative character of the natural world is supported by contemporary insights from cognitive science, particularly those theories that assert thought extends beyond the human organism and into the environment. The semiotic case for communication within nature helps position the idea that thought, cognition, occurs beyond just the human organism.

Does nature think, or at least contribute to thought? Consider first a person's work environment. In an office, the world of work is filled with tools that help us think through and solve problems: notepads, pens, networked smartphones and computers. Effective thought relies on such cognitive prosthetics. Expert professionals are so dependent on tools to aid their thinking that it is difficult to imagine undertaking any demanding cognitive task without some aid or the other. We used to think that a designer at least needed a pencil in hand to think through a design task. Nowadays such thinking tools include drawing and computer-aided design (CAD) systems, smartphones and browsers that access online image repositories. This much is obvious.

As these thinking tools become more sophisticated it gets easier to attribute some of that cognitive effort to the tools themselves. Therefore, CAD systems include libraries of predesigned parametric elements; paint programmes include palettes of colours and textures; and a scholar writing an essay draws on the work of others via books and articles delivered through search engines. Most of us for much of the time still preserve the idea that there is a person (agent), or possibly a group, who controls and takes responsibility for the cognitive task at hand, and there are methods for appropriately acknowledging other people, devices and systems that helped us on the way. But, thanks to networked computers, it is also easy to adopt the view that there is an aspect of cognition that is distributed. Some theorists then expand this observation into the idea that it is just as sensible to consider the entity that thinks, the mind, as also distributed.

It is worth considering not just computer tools but everyday tools and environments, and not just professionals who solve difficult problems but the rest of us caught up in everyday, mundane cognitive tasks such as finding our way from the kitchen to the living room, looking at the clouds to assess the likelihood of rain or speaking and making ourselves understood. Philosophers Robert Wilson and Andy Clark provide a helpful summary of the arguments for and against the idea that mind extends into the environment. They say of situated cognition, embodied cognition and extended mind: 'One way or another, all these locutions aim to suggest that the mind and the cognitive processes that constitute it extend beyond the boundary of the skin of the individual agent.'[64] Wilson and Clark seek to show that 'thinking is a kind of building, a kind of intellectual niche construction that appropriates and integrates material resources around one into pre-existing cognitive structures.'[65] Their model couples people's personal cognitive apparatus (brain and body) to the environment. The environment influences the way we think, and even what we think, and architects and designers have long believed that the environments they design influence thoughts, feelings and actions. However, the idea of extended cognition goes further.

I was drawn to this particular article as Wilson and Clark frequently use the terms 'nature' and 'natural'. They assert that

> natural resources, including cognitive resources, can simply be used by organisms, but sometimes this use does not merely fuel a pre-existing system – as in the above-mentioned cases of the respiratory and digestive systems – but augment the system itself and the capacities it possesses.[66]

The environment it is in augments an organism's cognitive capability. On the subject of biological evolution they re-quote an earlier article in which they asserted:

> Evolved creatures will neither store nor process information in costly ways when they can use the structure of the environment and their operations upon it as a convenient stand-in for the information processing operations concerned.[67]

They then focus (in this article) on the artificial, high-tech socio-cultural world of sophisticated human being.

But the case for extended cognition is strong even if we appeal only to the unadorned low-tech world of the human in the wild.[68] I will present the case in Chapter 8 that animals and our relationships with them furnish us with sophisticated cognitive constructs such as classification, boundaries, transitions and a sense of the other. In this and other respects, nature features as the primary thinking place. That is as useful a definition as any: nature as a place to think with. Following the biosemiotic thesis we could also say that nature is where thinking takes place. Forests think.

What is the place of data in this world of distributed cognition? A few times I have mentioned that from the position of semiotics, data is something built on the communicative actions of so many agents, natural and artificial. Data does not underpin these phenomena, but presents a derived implementation as a series of representational formalisms to some end, namely computation. This view resonates with Heidegger's concept of being-in-the-world. Heidegger constructs an ontological hierarchy at odds with the DKIW model of information science. His phenomenology positions bare facts, sense data and the world of objects as derived understandings, built on a world of experience.[69] Facts and sense data provide the most rarefied and abstract understanding of the world that is possible.[70] This is the realm of the self-sufficient subject engaged in pure contemplation or perhaps undirected curiosity. It is the most elusive and fragile encounter with the world. It requires appropriating without purposes or prejudices the supposed materialness of an object world. I think big data provides such encounters, especially when its advocates, such as Anderson, argue that data is beyond theory, and obviates the need for theory.[71] It is this final derivative and decontextualized level of being that is the basis of the Cartesian (traditional) ontology to which Heidegger directs his objections. Heidegger's phenomenology involves a fundamental 'reversal' of the understanding provided through the Cartesian ontology. It is worth noting that Heidegger does not see semiotics (signs) as a fundamental basis for human understanding of the world, but he does place it in a category of relationships in the world that is more ontologically important than data.[72]

Data is not a primary constituent of the natural world, but what of code? Having presented the case for semiotics over data as a way of understanding the natural world, I now turn to the challenge of disentangling the imperatives linking cellular DNA to the idea of computer code.

CHAPTER 4
BIOHACKING

Biohacking is simply amateur biological science conducted in a modest, low-cost environment, but by deploying sophisticated data and equipment formerly within the reach only of large companies and research institutes. Biohacking is a reaction among scientists and amateur scientists against pharmaceutical companies, genomics and other expensive bioscience research and development (R&D).

In this chapter I will examine some of the cultural impacts of the information-processing model of the gene. As ideas about data and code permeate biology, they embolden some scholars towards grass-roots innovation and democratization in the study of nature. They adopt the slogan of digital activists: 'Information wants to be free.' I examine the code metaphor critically, before turning to the idea of the meme, reviewing how certain biological metaphors permeate cultural transmission, how slogans gain currency and *memes* happen. My discussion therefore highlights a further emerging codependency between nature, semiotics and digital culture.

Garage biology

The antagonists against whom biohackers set themselves include large biotech and pharmaceutical companies, as well as peer review and regulation. The enemy is 'BigBio'. There is an interesting spatial dimension to biohacking activity in terms of where it happens. In his book on the subject, Alesandro Delfanti asserts:

> This so-called 'garage' or 'citizen' biology is conducted in weird places such as garages or kitchens and ranges from high-school-level educational experiments to complex biotechnology projects put into place outside institutional settings such as university or corporate laboratories.[1]

Since the sequencing of the first human genome in 2001, DNA patterns are now easier to discover, record and circulate. Working with DNA sequences also provides access to biological analysis without getting material on your fingers. You can play about with patterns, without having to deal with chemicals and wet matter. This is a culture where 'informational pattern is privileged over materiality'.[2] Though the biohacking movement is anti-establishment, and carries obvious risks as a 'danger to public health',[3] Delfanti points to biohacking as a return to the unregulated Victorian 'gentleman scientist' of independent means. After all, Charles Darwin started out as an independent hacker of sorts, working from home, and without initial institutional backing.

Delfanti notes soberly that innovations brought about by biolabs are not yet of consequence. Genuine biolab achievements include extracting DNA (e.g. from strawberries), basic DNA cloning and modifying the DNA in certain bacteria to cause them to glow in the dark. According to Delfanti, 'Right now citizen biology is not a site of research and innovation but rather of political, artistic and educational experimentation.'[4] As with many grass-roots innovations, large corporations are still involved. By encouraging and tapping into biohacker culture, some biotech companies and institutions engage with the creative potential of their publics. Here, biotechnology intersects with enthusiasts of fan-fiction, video gaming, and other creative and risky creative experiments 'in the wild'.[5] Many publishers, game developers and others in the creative industries have learnt not to resist such sources of enthusiasm and innovation, but harness them as a means of building audiences and fostering innovation.

Delfanti shows how large companies such as Google operate with the mindset of the entrepreneurial open source hacker, or at least that is part of their brand image. The Google spin-out company 23andMe provides a service for analysing DNA. The '23' in the name refers to the 23 chromosome pairs in human cell nuclei. The 23andMe website states that after submitting a saliva sample by post you will 'view reports on over 100 health conditions and traits; find out about your inherited risk factors and how you might respond to certain medications; and discover your lineage and find DNA relatives'.[6] Not only do you find out about yourself from this service, but you also contribute to a pool of data to be circulated and used by researchers.

Biohacking illustrates how deeply entrenched the metaphors of data and code are, deployed here to resist the power structures of biotech industries. The movement is inspired by open source developments in computing, driven also by the assumption that DNA operates as a kind of code. Hence 'sharing of genomic data through open access databases, the cracking of DNA codes'[7] and 'cracking nature's secrets'[8] go hand in hand with sharing software and equipment for home-based, low-cost bioDIY. The idea that nature contains codes is powerful, and DNA provides a strong source of evidence for this. It is worth examining the code metaphor more closely. After all, code here conjoins digital networks with nature.

Code

A code is a system of rules for translating one string of symbols (e.g. 1s and 0s) into something else, such as another string, some actions or outputs. Considering the discussion in Chapter 3, think of these strings as data. As indicated in the above account of biohacking, code provides an obvious link between digital computers and nature. After all, computers run on code, and the nuclei of cells in living organisms contain gene sequences that the cells translate to other sequences, proteins and actions, and those translations imply the actions of a code. In advancing the significance of code in living things in the 1990s, computer scientist Christopher Langton stated: 'Our technological capabilities have brought us to the point where we are on the verge of creating "living" artifacts.'[9] The field that resulted is 'Artificial Life', that 'is devoted to studying the scientific, technological, artistic, philosophical, and social implications of such an accomplishment'.[10] Code is of consequence in the relationship between machines and nature.

The code metaphor is strong in biology. A section headed *The genetic code* in a biology textbook by William Keeton states,

> We are dealing with a code that has only four elements – the four different nucleotides in messenger RNA (which reflect a corresponding four nucleotides in DNA) … nucleic acids must code an immense amount of information within a small space … the flow of information proceeds as follows: The DNA of the gene determines the messenger RNA, which determines protein enzymes, which control chemical reactions, which produce the characteristics of the organism.[11]

I am less concerned here about the details of the process described than about the frequency with which the word 'code' appears, as if assumed, and its company with the word 'information'. The explanation draws substantially on the terminology of individual units, code and translation.

Biologist Richard Dawkins provides further elaboration of how genes steer the development of an animal embryo.[12] As embryos develop in an orderly way, Dawkins thinks of genes as a complicated recipe for the organism's development, 'like the procedure for making a cake, except that there are millions more steps in the process and different steps are going on simultaneously in many different parts of the "dish."'[13] Cells duplicate, multiply, die and join to form tissue and multicell structures, including organs. Only a small number of the genes are active in any part of the developing body, and at any particular moment: 'Precisely which genes are switched on in any one cell at any one time depends on chemical conditions in that cell. This, in turn, depends upon past conditions in that part of the embryo.'[14]

The complete set of DNA material is the same in all cells of any particular organism, and is called the genome. It happens that the human genome sequence for a typical human is displayed in a series of volumes as an art installation in the Wellcome Trust gallery in Euston Road, London. The gene sequence is about 3,000 million characters long if the contents of a human cell's 23 chromosomes are strung together.

Is DNA a code in the same sense that we speak of computer code – and is the cell a computer? Philosopher and historian of science Lily Kay thinks biology has been led astray by the code metaphor. She maintains that 'the genetic code is not a code: it is simply a table of correlations'.[15] She likens this genetic tabularization to the periodic table, though it is not nearly as effective at systematically predicting chemical behaviours. The unpredictability of gene sequences comes about 'because of contingencies, degeneracies, and ambiguities in the structure of the so-called genetic code'.[16] The code metaphor will not rest, however. It is worth pursuing further the hold it has over nature.

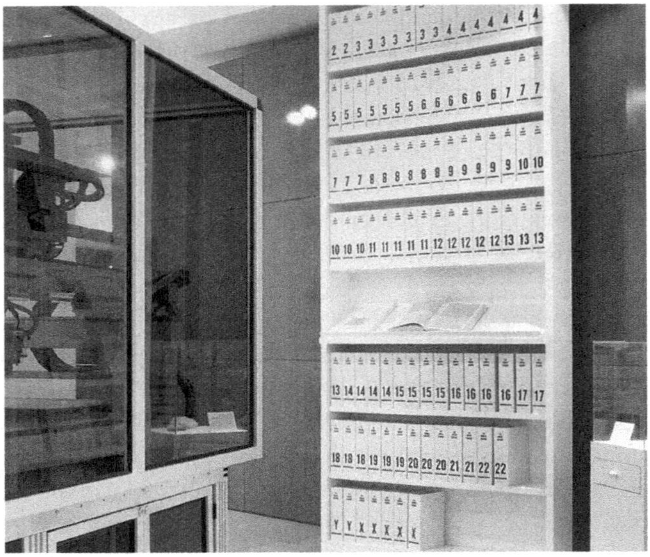

Figure 4.1 Books containing the human gene sequence at the Wellcome Trust, London.
Source: By Russ London at English Wikipedia, CC BY-SA 3.0, https://commons.wikimedia.org/w/index.php?curid=9923576.

Life is not code

Alan Turing is widely regarded as the founder of modern computing.[17] His generalized model of computation shows that in a computer there is no substantial difference between data and programs. So, 'code' will do for anything processed in a computer (data or instructions).[18] A Turing machine is a hypothetical computer stripped down to basic operations involving a tape of indeterminate length with symbols on it, a tape head for reading and writing symbols, a mechanism for moving the tape back and forth, a register for recording the machine's last move, and a finite set of instructions.[19] Were it to be built, such a device would be slow, inefficient and impractical, but has properties that account for everything that a computer can do, including mathematical operations, and text and symbol processing. All computers would be Turing machines were it not that they lack the luxury of anything resembling an infinitely long tape and endless storage capacity. However, a computer does not need the elements of a Turing machine to be a subspecies of Turing machine. Modern computers process their data whether it is on tape, disk or hardware much more efficiently than moving a reader back and forth over a string of symbols.

But the form and structure of the hypothetical Turing machine suggests interesting links to biology. It inspired biotechnologists Ehud Shapiro and Yaakov Benenson to draw on the similarity between a Turing machine and the way a cell and its genome function. They observe:

> Both systems process information stored in a string of symbols taken from a fixed alphabet, and both operate by moving step by step along those strings, modifying or adding symbols according to a given set of rules.[20]

Though a computer does not need to have the same configuration of elements as a Turing machine does, Shapiro and Benenson are encouraged by these parallels to speculate that 'biological molecules could one day become the raw material of a new computer species'.[21] If the cell, and its genome, is a Turing machine then the genome and its cell can be used to perform calculations. The challenge focuses on how to turn cellular processes into useful computers. The chemical processes in a cell are much slower than in a computer, and there are many constraints, but one of the payoffs of such biological computing would be that: 'Tapping the computing power of biological molecules gives rise to tiny machines that can speak directly to living cells.'[22]

Getting cells to do computation is not so farfetched. After all, we human beings can do calculations, and we are ostensibly made up of cells. The reverse is more of a challenge: can computation be made to do what cells do? As known to many working in the field, the strength of genomics and genetic engineering resides in understanding, manipulating and even controlling biomaterial substances, not fabricating biological material out of data or information in a computer. The a, g, t and c symbols used conventionally to indicate the elements of the gene sequence in the genome stand for molecules, each with individual properties influenced by their surrounding conditions. They also exhibit emergent properties in combination. I need hardly say that the symbols themselves do not deliver material properties.

Codes work on symbols, which might eventually control the moving parts of a machine, robot or other actuated synthetic entity. Building on the critique of the code metaphor by Lily Kay,[23] at most, the process of translating and mapping between symbols might lead eventually to a simulation of chemical behaviours, or even an aspect of the behaviour of the cell or organism. But no amount of substituting, translating, converting symbols or simulation will result in an organism. For that at least the actual materials, the chemicals, are needed. Data are not atoms. The books of character sequences in the genome exhibition in the Wellcome Trust gallery remind us that sequences of symbols are maddeningly lifeless. There is no life in those volumes. Biomechanists have to look elsewhere to support the proposition that computers and nature meet at the micro-scale, that we are at the brink of a post-human merging between machine and organism,[24] or that machines will assume vital characteristics of organic life, such as agency, inheritance, self-reproduction and genetic evolution across generations.

Cellular semiotics

As I have shown, biosemiotics is an area of research that studies living matter in terms of its communication networks, that is, the transfer of signs. Biosemioticians think that the exchange of signs applies to whole organisms communicating with one another (as when humans or other animals communicate). They also think the exchange of signs applies in the microscopic and chemical domain, including the way cells process DNA. Signs require interpretation, and it is here that biosemiotics brushes against theories in hermeneutics. Anton Markoš and his colleagues explore biology from the point of view of semiotics and hermeneutics. Their work draws on relevant writings by Martin Heidegger and Hans-Georg Gadamer.[25] Markoš takes issue with the conventional code-oriented, culinary view of DNA processes, that is, Dawkin's recipe metaphor.

Genetic instructions are complicated. Markoš thinks that the 'objective of hermeneutic biology should be to get rid of the genocentric view that takes the genome as a recipe for building the body'.[26] He also extends Heidegger's view of who and what is entitled to come under the category of the living and communicative being who is 'in-the-world'.[27] Whereas Heidegger restricted his concerns to those entities that participate in spoken and written language, Markoš follows the biosemiotician's line that includes all of life and all living creatures as language practitioners.[28] He says, 'all living beings relate themselves to the world through "languageness"'.[29] The word 'languageness' suggests a primary kind of language that precedes what is spoken and written – a protolanguage. It is important for Markoš to extend language to the unspoken as he wants to describe what happens inside cells in terms of language. He thinks of a cell as an interpreter or reader of DNA. It is worth quoting an extended passage from his book:

> The string of DNA is a genuine text that is read and interpreted by the cell. The cell will behave according to how it interprets the text: the interpretation is based on previous understanding, momentary contexts, and/or layout for the future. This metaphor is again in opposition to the metaphor of the genetic script as a program or an algorithm. A text written in natural language is not an algorithm: it requires a reader, not hardware.[30]

Here he echoes a standard hermeneutical account of how a human reader engages with (i.e. interprets) a text. One of the threads he develops in support of his argument notes the presence of redundancy in the text that is the genome. Much of the DNA sequence in a genome is superfluous, or it just holds the strand together, or switches on or off as the context changes or it serves as backup if other strands fail due to mutation. Markoš thinks that is a lot like spoken language, in which there are many substitutes for a word. For example, the word 'dog' can refer to the animal in general, the male animal, an insult, a creature that is crude, difficult or insistent, and 'dog' has many grammatical forms and substitutes (doggy, dogged, mongrel, hound, pooch, puppy) that offer different meanings depending on context. Words in spoken and written language shift and change and require interpretation, as does the whole passage, book or corpus in which the words occur. So too, Markoš asserts that the words of a DNA strand require interpretation in terms of their place within the bigger context of the organism, population and environment.

Markoš also invokes the Heideggerian concept of *tuning*, to which I referred in the Introduction, to describe how a cell operates. The interpretation of the cell's DNA depends 'on the quality of the text itself and the "tuning" of the cell (the above-mentioned coordinates, physiology, morphology, and history) and is indeed to be viewed as a hermeneutic task'.[31] In the midst of an extended discussion invoking technical terms such as 'nonlocal morphogenetic fields', 'concentration gradients', 'an extracellular matrix', 'cytoskeleton' and 'nucleoskeleton' Markoš asserts: 'DNA is thus far from being the algorithm prescribing how the body will look and how it will behave. It is a genuine text to be read by an informed (or better initiated) reader'.[32]

He thinks the timescale of such interpretative operations varies: 'Changes in the interpretation of any of this multilayered information accumulated over billions of years can also be considered mutations and may even result in misinterpretations ending in aberrant development, or tumors, but from time to time also to a new morphological variant'.[33] Markoš's argument treats the cell as an interpretive agent. This attribution of the capacity to interpret beyond the limits of human agency engaged in spoken language provides a provocative departure not only from genetics but also from mainstream hermeneutical study. It is useful to think that interpretation works all the way through the organism, from the thinking human in society down to the cells of which a person is composed, and spans across all living things. This is part of the message of biosemiotics, that the exchange of signs, verbal and otherwise, pass through the entirety of living systems. In fact, semiotics suffices without insisting that cells use language. As I have already indicated, Peirce and others see language as a subspecies of the more general class of semiotics, the communication of signs.

One might criticize the comingling of divergent fields such as biology, semiotics and phenomenology in this way, and transferring terms such as interpretation, translation and communication from one field to the other.[34] But I think the ambition is useful as it contributes to a raft of arguments that support the unity of natural and cultural processes.

Cultural memes

There are other impacts from the DNA as code metaphor, namely the extension of the processes by which genetic material is duplicated, preserved, mutated and transmitted. That is the story of the *meme*, bringing DNA into social media, architecture, nature and other cultural discourses. *Memes* are like genes in the social realm, according to the main proponent of the idea, Richard Dawkins. He says: 'They [memes] are patterns of information that can thrive only in brains or the artificially manufactured products of brains – books, computers, and so on. … As they propagate they can change – mutate'.[35] Memes also jump (unlike genes) from one replicator medium to another,[36] for example, from brains, to books, to computers and back again. Following meme logic, humans are (just) machines, among other replicator machines, for propagating memes and for

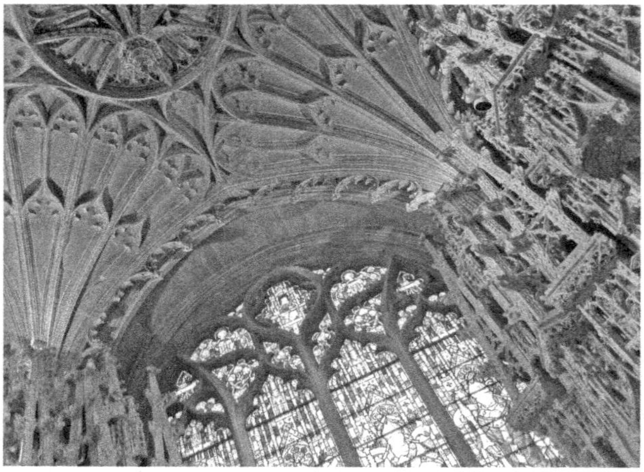

Figure 4.2 Fan vaulting as a meme, Ely Cathedral.
Source: author.

effecting the memes' transition from one replicator medium to another, according to psychologist Susan Blackmore in *The Meme Machine*.[37]

As with DNA, the meme idea has its origins in the concept of code, where information is delivered overtly as sequential patterns. Spoken and written languages are the most obvious examples of media for the transmission of memes. The usual meme vehicles are texts, as they appear in books, libraries, databases and social media exchanges. In this constellation of media, meme theorists place human beings on a level as one type of meme carrier among many, as biological machines, or simply 'brains'. Memes can also be material and visible other than through text. Meme vehicles include buildings, their components and other artificial constructions. Some artists, designers, architects, and engineers think of textual production as just one category among a range of potential meme carriers.

According to John James,[38] historian of Gothic architecture: 'A meme is like a catchy tune, a new fashion in clothes or a way of building an arch. When an architect hears about a good idea, he passes it on to his colleagues and students, and if it catches on it will, like a gene, propagate itself by spreading from brain to brain.'[39] A gothic arch, or a fan vault, is a meme passed on through the generations of church construction practice, modified and adapted over time, like a gene.[40] Architectural theorist Chris Abel concurs with the meme hypothesis, while drawing attention to the meme's network ecology. He agrees that technologies, including buildings, are carriers of memes: 'All buildings and other artifacts are embodied technical memes of one kind or another.'[41] Drawing on Manuel de Landa's (and Deleuze and Guattari's) concept of *assemblages*, Abel also argues that 'no technical meme exists on its own, but always belongs to some larger combination or series'.[42] If we are permitted the neologism, *archi-meme*, then the meme does have currency in architectural and built environment discourse, amplifying further a kind of ecosystem of forms and ideas in the built environment, complete with processes on a par with the code of life itself.

Social media memes

The meme idea has gained renewed currency in contemporary discussions involving digital communications and social media. Microblog postings such as tweets are a bit like genes in that they get reproduced many times over. They mutate, thrive and persist if the environment is right. As for genetic material, there is a lot of redundancy and wastage. Most tweets do not get retweeted, liked or commented on. So, that is something like natural selection.

One of the provocative ideas about genes is that humans and other organisms (the phenotypes) are the vehicles by which genes transmit themselves. According to Richard Dawkins the genes are 'selfish' agents intent on their own survival and multiplication.[43] The genes that produce phenotypes with the best characteristics for survival in the phenotype's environment are the genes that survive and pass their DNA on to subsequent generations of phenotypes. We complicated and elaborate humans are (just) machines for the replication and transmission of genes, that is, DNA molecules. That proposition appeals to post-humanists intent on counteracting the conceit that humans are the only beings in control of nature, or that matter, in the world we live in.[44]

Microblogs as memes sound plausible as a metaphor. Tweets are patterns of information after all. So perhaps they are memes, or contain memes.[45] The idea that we human beings are replicators of elements in language is not so alien. The autonomy of language is an idea that finds support from the philosophical proposition that we human interlocutors cannot escape language, and that rich, meaningful and engaging conversations have their own momentum. According to Gadamer: 'The way in which one word follows another, with the conversation taking its own turnings and reaching its own conclusion, may well be conducted in some way, but the people conversing are far less the leaders of it than the led.'[46] The idea of the meme illustrates further how data-oriented biological discourse infiltrates our understanding of human culture and our world view. Biology provides a rich source of metaphors, and like any metaphor enables certain possibilities and occludes others.

Biosemiotic discourse

As I am writing about language and nature, it is worth drawing this chapter to a close by reflecting on how we talk about nature, as a meta-discourse, that is, the semiotics of the discourse itself. In Biology 101, amid formaldehyde-soaked frogs and the collective scalps of snap-frozen drosophila, we learnt the language of nature: zygotes and gametes, monocots and dicots, dominants and recessives, liverworts and mosses. But the zoo of terms has expanded. Scholars and practitioners of biomimetic art, architecture, engineering and computing have appropriated and invented a slew of terms grafting nature to their discipline. These are key elements in the network of biological memes.

Rachel Armstrong provides a helpful summary of terms in her book: *Vibrant Architecture*.[47] She defines 'vibrant architecture' as 'a stochastic form of architecture that is compatible with a Nature-based method of architectural production'.[48] Vibrant architecture is less about improved industrial practices based on nature's models, buildings as living entities, or artefacts inspired by the natural world; rather, it is an architecture that taps into nature as 'deconstructed and stripped of its aestheticisms to reveal its raw, relentlessly material character'.[49] I interpret this crudely as an architecture that acknowledges the informal, and the compositionally inconvenient, if not inelegant, nature as mess. Other terms she defines include: *process philosophy, complexity, actant, assemblage, natural* and *post-natural.*[50] But none of these terms have been as virulent and invasive as *biology* itself.

Commonly occurring bio-oriented portmanteau terms I have encountered in the bio-architectural literature include: *bio-3D-printing, biodesign, bioDIY, biolinguistics, biomimesis, biomimicry, biomorphic,* and *biotechnology.* Further terms populate the *Routledge Companion to Biology in Art and Architecture.*[51] As my subject is semiotics, it cannot go unremarked just how amenable certain terms are to assemblage, and subsequent proprietorial and disciplinary claim, branding and even innovation. The prefix *bio-* derives from the Greek *bios*, meaning not just *life*, but *course of life*. A *logia* is a discourse. So, a biology is a discourse about the course of a human life, that is, what happens, or at least an account of what happens, during someone's life. Biology is the study of the life course, and not just of humans but any organism.

One pleasing outcome of this etymological excursus is that biology is grounded in storytelling, language and communication. Biology enjoys this literary and even semiotic foundation of sorts, at least as expressed in English and other European languages that draw on Ancient Greek and Latin. The classical (Greek and Latin) words *geographia, genesis, organum* and *technologia* provide similar fecundity in the lexicon of linguistic and disciplinary invention (*geo-*, *geno-*, *organo-* and *techno-*).[52]

In exploring the 'gene as code' metaphor I reviewed how ideas about data and code permeate biology. Some scholars think the code metaphor points to grass-roots innovation and democratization in the study of nature. I examined the code metaphor critically. Certain biological metaphors permeate cultural transmission. I come down on the side of the materiality of genetic substance identified and described through various metaphors. The code metaphor has utility, but is limited, especially as we think about the materiality of life. I also looked at the meme, and the cultural transmission of ideas from biology. My discussion therefore highlighted a further codependency between nature and culture, expressed not least in the meme that we can 'crack nature's secrets'.

Already we can see that semiotics is biocentric, that is, centred on the organic. Organisms and their components communicate, transmit and pick up signals. But do rocks, gases, and planets communicate with one another? We turn sunrises and sunsets into symbols of the start and end of life, but does a red sky as symbol have anything to say to the rocks of the mountain on which it throws long shadows? Do the clouds, stones and caves on Mars (if we assume there is nothing organic there) have anything like names in the commerce of signs among themselves independently of human exploration? I will respond to these questions in subsequent discussions of agency, life, enchantment and numinous nature, and after a discussion about reproducing nature.

CHAPTER 5
REPRODUCING NATURE

The forms, structures and functions of leaves, flowers, insects, mosses and other elements in nature survive harsh competition. Why not copy them? In this chapter, I will review how bio-designers seek to copy the forms, structures and functions of nature. Some bio-designers also observe and learn from nature's processes in the production of artefacts, capitalizing on the assumed affinity between nature and digital code.

Natural artefacts

Connoisseurs of nature paintings explain such mimetic motivations in aesthetic terms, such as the human tendency to admire the beautiful and the sublime.[1] Some psychologists refer to *biophilia*, a putatively innate affinity with things natural that encourages us to mimic living nature.[2]

Viewed through the lens of semiotics, the reproduction of nature presents another example of sign systems in operation. In accordance with Sebeok's classification of signs discussed in Chapter 3, a natural thing in the landscape such as a tree in a desert can *signal* to animals and human travellers the presence of water. Dead leaves are *symptoms* of excessive groundwater. The foliage is an *icon* of the volume of rainfall (the greenness and fullness of the leaves represents in some way the amount of rainfall). The state of the tree communicates the presence of nutrients in the manner of an *index*. Changes in leaf colour operate as *symbols* of the changing seasons, and trees communicate distinctions in structure, lineage, form, movement and scent that place them in easily detected categories as identifiers, that is, as *names*.

The *reproduction* of a tree in a drawing, painting, photograph, film or digital model also functions as a sign. Representational paintings have an iconic relationship with their subject matter, and a conventional photograph serves as an index of a tree by virtue of its causal technical processes.[3] I will here forgo the task of cataloguing examples of art works and their functioning as signs under Sebeok's system. In any case, to identify examples of signs is an exercise in interpretation, and the choice of category depends on the context of the discussion. Signs are varied, and the language of semiotics provides a vocabulary for thinking about different aspects of communication. Cataloguing sign functions can be an abstruse task, and Sebeok's system is simpler even than Peirce's system from which it is derived. However, I feel it is worth pursuing these semiotic narratives as they help situate architectural and digital reproductions of nature within an ecology of signs, and also help to make sense of the natural–artificial problematic.

Architecture presents many opportunities for semiotic analysis, not least as it deploys elements from nature in ornamentation, wall paintings, in its use of materials and the forms and shapes of its elements. Prominent among architecture's sign functioning is its attempt to mimic nature, or at least there is a propensity among some architects and engineers to claim they are copying the processes by which forms, shapes, colours and functions in nature develop. In his book titled appropriately *Iconic Building* Charles Jencks draws on the terminology of semiotics:

> An icon (*eikon*) is literally a 'likeness, image, or similitude', such as that of a saint painted on a wooden devotional panel. The word always carries this old religious meaning as does its negative, iconoclasm.[4]

Much iconic architecture mimics nature, natural forms, elements of nature or natural processes. Sometimes biomimetic architecture stands out as spectacular therefore, demanding what Jencks identifies as the kind of awe and veneration of religious imagery, still named as *icons* in Catholic and Orthodox practices.

Of course, iconic architecture can also mimic things that are not of nature, including machines, other buildings, people, ships and consumer products, and the similitudes on which architecture draws can move from the tangible and obvious to something like an ideal, a spirit of the times, or a rupture.

Some of Jencks's own landscape architectural interventions aim to represent the nature of the universe as revealed through chaos theory and quantum mechanics,[5] by replicating the smooth growth of wave forms in nature and their sudden twists and ruptures: 'Since nature shows these two properties might not a cosmogenic architecture represent these two basic truths?' Hence, he is attracted to the fluid forms of Frank Gehry's Bilbao museum. Such biomimetic architecture also bears the derision of architectural theorists such as Alberto Perez-Gomez in his phenomenological account of architecture: 'Regardless of very diverse formal preferences ranging from orthogonal to organic shapes, crystalline to folded, simple or complex "tectonics," the aim is mostly that of effect.'[6] For such critics, semiotics deals simply in how things appear, and hence deals in superficialities.

I mentioned the semiotics of trees above. Trees feature prominently as sources of architectural biomimesis. Classical fluted columns are stylized tree trunks. Tour guides describe Enrique Morales's design for the Scottish Parliament building as resembling 'the tree of life' when viewed from Salisbury Crags. There is also Pier Luigi Nervi's high-modernist structures, and expressionist and organic architectures earlier in the twentieth century, as well as Pugin's Gothic revival architecture; and of course, Vitruvius thought of architecture as deriving from the manipulation of a copse of trees into a shelter: four trunks define the corners and the canopy becomes the roof to make the primitive hut. By his account human kind emerged from the forest and developed a sense of community around the fire.

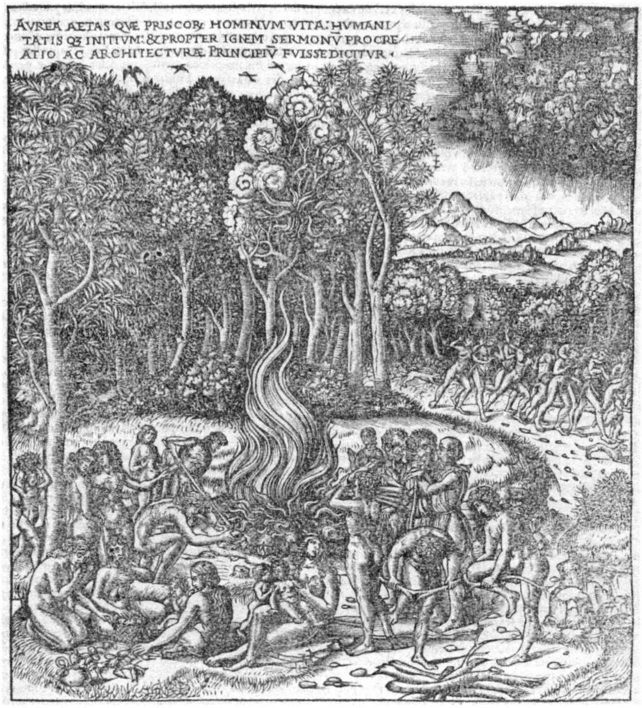

Figure 5.1 The discovery of fire. Illustration by Cesare Cesariano (1475–1543) to Vitruvius's *Ten Books of Architecture*.
Source: Warburg Institute, public domain.

Architectural commentator Sarah Williams Goldhagen claims that developments in cognitive neuroscience also affect architecture, not least in influencing the forms of buildings. She refers to the redevelopment of the Plaza de la Encarnacion in Seville, Spain, which is a spectacular series of computer numerical controlled (CNC) forms suggestive of rock formations, caves, clouds and tree canopies.[7] She also references arboreal structures by Zaha Hadid and others. Lest we think that all references to nature are adulatory, trees can also carry negative connotations. For example, Deleuze and Guattari deride 'arboreal thinking', which they regard as hierarchical,[8] and Christopher Alexander declared: 'A city is not a tree.'[9] Whether deliberately or inadvertently, a sign can operate in diverse ways.

Most often, biophilic design is design that purports to be sympathetic to nature, which I take to mean it participates in a semiotic ecology with nature as its focus. Designers who want their buildings and landscapes to exhibit biophilic qualities have several attributes to draw on. Social ecologist Stephen Kellert in the book *Biophilic Design* lists several: use natural colours, water, plant motifs, natural shapes and forms (like trees and shells). The design should allude to growth and other natural processes, introduce natural and filtered light, connect with history and the spirit of a place, promote curiosity, attachment and sometimes be daring. The list includes practically every good design practice. Perhaps biophilic design is simply good design.

As heirs to romanticism, many designers and critics champion wholeness, unity and coherence as the pivotal virtues of designing. Kellert asserts: 'People prefer in natural and built environments the feeling that discrete parts comprise an overall whole, particularly when the whole is an emergent property consisting of more than the sum of the individual parts.'[10] He adds: 'This integrative quality fosters a feeling of structural integrity, even in complexes of considerable size and detail.'[11] For such bio-romanticism, materials also ought to yield to the effects of age, change and the patina of time, a characteristic of natural materials. By way of contrast, 'artificial products rarely evoke sustained positive response even when they are exact copies'.[12] The book provides helpful images of iconic buildings and places that demonstrate biophilic harmony: the Sydney Opera House, 30 St Mary Axe (the gherkin) in London, ivy-covered courtyards and examples less familiar. Buildings do not need to be exact, or nearly exact, copies of nature, to fulfil biophilic criteria it seems. Their function as icons of nature can be interpreted very broadly, and it is fair to say that not all icons are interpreted and received with approbation, nor by everyone.

Digital biomimetics

Designers and artists may seek to reproduce nature, but what is the role of digital code in this creative ecology? Digital technologies provide the means of simulating the processes and forms of biology. Let us start with windows.

In her book *Technobiophilia: Nature and Cyberspace*, Sue Thomas captures the codependence of the technological and the natural, elaborating on the role of digital technologies in expanding our affinity with nature.[13] She makes no direct reference to semiotics or signs, but her example of a window provides a telling example of the way digital technologies intersect with biology. Windows have an important role in bringing the supposed therapeutic and restorative benefits of natural environments into buildings. Thomas notes the value of windows as a means of enhanced contact with 'nearby nature'. Windows are one of the means by which such signs are transmitted from nature. They are a medium of communication. Being able to see, hear and otherwise sense the world on the other side of a window provides a crucial communicative channel. Thomas is not interested in plate glass windows with barely visible frames merging inside and outside, but windows that render conspicuous the threshold between the world within and what we choose to identify as the natural without – trees, grass, clouds and sparrows. Her archetypical window experience is that of the home-working academic with desk and laptop affording glances out through the deeply set window of a thick-walled country cottage. Calling on insights from architect Kent Bloomer she advocates thickening the window threshold with patterned curtains or blinds, decals and window ornaments.[14] Frames, mullions, astragals, sills and deep reveals enhance the boundary condition. In semiotic terms, this thickening renders the window conspicuous as a sign, rather than treating it as an invisible threshold.

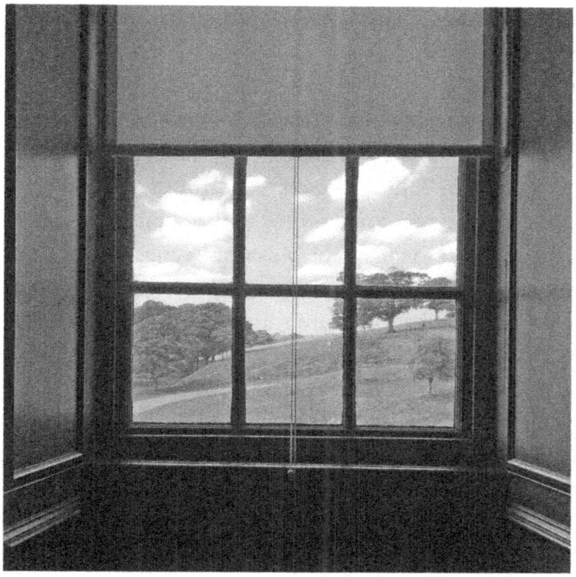

Figure 5.2 Deep window reveal emphasizes the relationship between inside and outside. Lyme Park, Disley, Stockport, England.
Source: author.

Following up on Bloomer's article I see that he cites Frank Lloyd Wright's ornamental window treatments. Bloomer states that 'the moment of divide is the most charged, ambivalent, and negotiable for belonging to both sides of the psychological boundary that informs our reaction to the environment'.[15] He argues that as a poet creates textures from words, so windows and other conspicuous architectural elements coax buildings towards the poetic.

Before examining what Thomas says about 'digital windows', it is worth reviewing the window as a sign. A window as sign operates according to the six sign types as outlined by Sebeok. An open window can *signal* that the climate outside is sufficiently benign to admit the communication of air between the inside and the outside. A window dripping with moisture is a *symptom* of temperature and moisture variation. A window is an *icon* in so far as it functions as a picture frame. For inhabitants of a building a window bears an *indexical* relationship with the outside and points inevitably to elements in the view beyond.[16] The window is also a *symbol* and references ideas about connection, access and transparency, and it serves as an identifier or *name* that stands in place of what is outside. In these cases, here I have also qualified the window as open, wet, framing and as providing access. It is not just the window that serves as a sign but its situation and the state it is in.

Windows provide a potent symbol of the relationship between architecture and nature. In *Technobiophilia* Thomas outlines Stephen and Rachel Kaplan's arguments for the benefits of being in natural environments, and translates these to the online world.[17] She draws on studies focusing on people's responses to the presence or absence of windows.

Apparently, hospital patients recover more quickly, and people in waiting rooms are less anxious, when there is a view of green space from a window. But the strong claim about exposure to nature is that pictures of nature scenes on the wall provide a similar benefit, as do still and video images of nature scenes: 'Whether indoors, outdoors or online, it is clear that nearby nature has a profound effect on well-being.'[18] She also refers to the book *Technological Nature* by Peter Kahn, who studied groups of people working in different environments: places with no windows, rooms with actual windows to an outside world, and wall-mounted computer displays of outdoor scenes simulating the view from a window. Kahn concludes, 'Technological nature is better than no nature but not as good as actual nature.'[19] Window substitutes are satisfactory if there is no better option. The idea of a window substitute on a wall is of interest from the viewpoint of semiotics. In so far as a window points causally to what is beyond it, the window functions as an index. A picture on a wall, an animated digital screen or other substitute window in this case serves as an icon. It bears a likeness to a window, though only in some respects: in that it is usually rectilinear, positioned on the same plane as a wall, and directs attention to something other than what is in the room.

But irrespective of the evidence about recovery and well-being, these authors are uneasy about the idea that we could fill our environments with window surrogates. They do not describe it in these terms, but it is evident to me that an occupant cannot approach a wall-mounted display screen and peer over the virtual sill to inspect the shrubs below. There is none of the depth alluded to by Bloomer, no 3D, parallax or other geometrical cues, let alone two-way communication with a world beyond. Nor does it accommodate William James's example of a lover tapping on a windowpane.[20] I hope I do not need to catalogue all the ways that an actual window differs from a picture or screen on a wall. I take from Thomas's *Technobiophilia* a challenge to the privilege we accord to a putatively authentic experience of nature. We are sometimes as content with icons as we are with indices. That similar benefits might be achieved by other (digital) means adds weight to this challenge. At best, we could say that windows, framed pictures, wall-mounted mirrors, display screens and flat-screen television sets exhibit complex semiotic relationships with one another. Of course, windows are caught up in a myriad of signs that include windows on computer screens, eyes as windows of the soul, books as windows into the world and other iconic associations between windows and other things they resemble. Such associational chains are in accord with Peirce's assertion that 'we have sign overlying sign.'[21]

These authors deal with our experience of nature, but not our experience of visual images. Looking through a window and looking at a picture of a scene through a window both involve interpretation, but referring to pictures of what we might see through a window invokes debate about the priority of indices over icons. Like a series of words in a book, a photographic image indexes the subject matter of the picture. In terms elaborated by another scholar of signs, Roland Barthes, a picture may *denote* a meadow but simultaneously *connote* a summer holiday, life on a farm, intensive agriculture, *The Sound of Music* or other images.[22] Some of these picture references may or may not impinge on our sense of well-being. Memories and culturally influenced preferences come into play. Above the fireplace in my dentist's waiting room is a gilt-framed print of a scene in the Swiss Alps. I do not think that reduces my anxiety about what may follow, but fortunately there is a large transomed Georgian window in the surgery room that turns my gaze towards some trees, and the possibility of freedom.

Digital skins

Apart from windows, another important element in the semiotic mindscape is skin. Skin has semiotic functions like those of a window. It is a communicative medium. The condition and appearance of skin *signals* something about the state of the organism. Its condition is frequently a *symptom* of an underlying pathology. The skin is an *icon* in that it bears the form of the animal that it encloses. When a medium of touch it bears an *indexical* relationship with the things to which it has made contact. It is a *symbol* of connection, sensitivity, identity and power, and it acts as a *name*, especially when it serves as an identifier of age, race, socio-economic status, and well-being, not least when adorned, powdered, covered, exposed, scarred and tattooed.

Digital screens, artificial sensors and actuators combine to form 'digital skins' as elements in buildings. These technically sophisticated icons serve to bind the connection between digital machines and nature. Living human and animal skin is palpably different from a touch-screen video display. Completely unadorned, raw, nature as you find it (e.g. unadorned human skin or a leaf), contrasts with the maximally manufactured, contrived, and artificial (e.g. a touch screen or a microchip). But the way people talk about digital technologies readily conflates the digital with the natural, or at least the organic. As an example, a book by Thorsten Klooster surveys the range of innovative 'smart' surfaces for potential use in architecture and design, referring to such concepts as 'evolutionary nanotechnologies' and 'biological synthetic nanotechnology'.[23] Researchers into such digitally honed surface technologies mine the organic, biological, living, sentient and natural for terminology, models and analogies.

In some cases, such artificial components incorporate themselves into the organic, and vice versa. Smart surfaces provide a good example of such natural–digital hybridity. These are the high-tech surfaces at various scales common enough as solar panels and smartphones with touch-sensitive display screens. They are beginning to scale up as architectural elements that define spatial envelopes, adjust environmental conditions, harvest and filter data from the environment and change the shape, function, quality and mood of a space. As known to every student of biology, much of the 'smartness' in biology resides in membranes, cell walls, tissues and skins, that is, surfaces. Synthetic smart surfaces do not need to wire up to digital circuitry but can include meticulously engineered pigments, coatings, fabrics and meshes made of tiny components designed to respond to environmental conditions through local bio-synthetic processes. Klooster describes a project to develop one such smart surface, referring to a 'nanostructured macro-surface … which functions quasi biologically'.[24] Such a surface 'is to be generated by growing nanotubes under the control of microengineering-based lithographic coating, printing and etching steps'.[25] The language here is an assemblage of engineering and biological terminology.

The biological and the digital hook up at the nanoscale. After all, biological processes are mostly of a size invisible to the human eye, as are microcircuits. The other commonality here between the natural and the digital resides in the use of the word 'system' to describe all natural and artificial processes.[26] For example, human skin is part of the human integumentary system, and the circulation of blood is part of the cardiovascular system. Several decades of research in systems theory seeks out common organizational substrates by which we understand biology, geology, human organization, infrastructures, machines, manufacturing and design.[27] Klooster's smart surfaces are after all systems (e.g. media facade systems, hybrid energy–generation systems, climate-control systems). Digital technologies are currently at the top of a hierarchy in which designers, engineers and architects have maximum scope to manufacture, adjust, simulate and test systems. Digital circuitry and its code provide boundless plastic media through which technology and artifice assert their putative mastery over the natural. From this point of view, a conflation of the biological and the digital seems *only natural*.

Within the semiotic framework, communication is possible through the working of signs. Sebeok says:

> The essential ingredient, or nutriment, of mind may well be information, but to acquire information about anything requires, via a long and complex chain of steps, the transmission of signs from the object of interest to the observer's central nervous system.[28]

Systems theory builds on the importance of such information flows. Mechanical, electronic, digital and natural systems are supported by the movement of information, and hence communication.

For Klooster and other biomimeticists, this apparently common substrate of information processing gives them further licence to conflate the natural and the digital. After all, information flows are manifested in the case of synthetic smart surfaces. Klooster describes surfaces by which people and machines communicate, that is, via interface technologies such as touch screens. There are also surfaces that serve as ambient displays that provide 'constant, situation-related, visible and accessible information'.[29] His taxonomy builds up to sensor arrays and networks of sensors passing information to actuators that modify surfaces and environments, and their nanoscale variants that might eventually manufacture, adapt and improve themselves as self-organizing systems, much as human and animal skin that repairs itself. Natural systems convey information (think of nerves), but they also convey energy, materials and actions. Some of this functionality is captured in the design of synthetic smart materials.

In so far as we give priority to code and information flows, the distinction between the natural and the digital appears arbitrary or perhaps irrelevant. More precisely, the natural is subsumed by code and data flows. Rather than dissolve the distinction

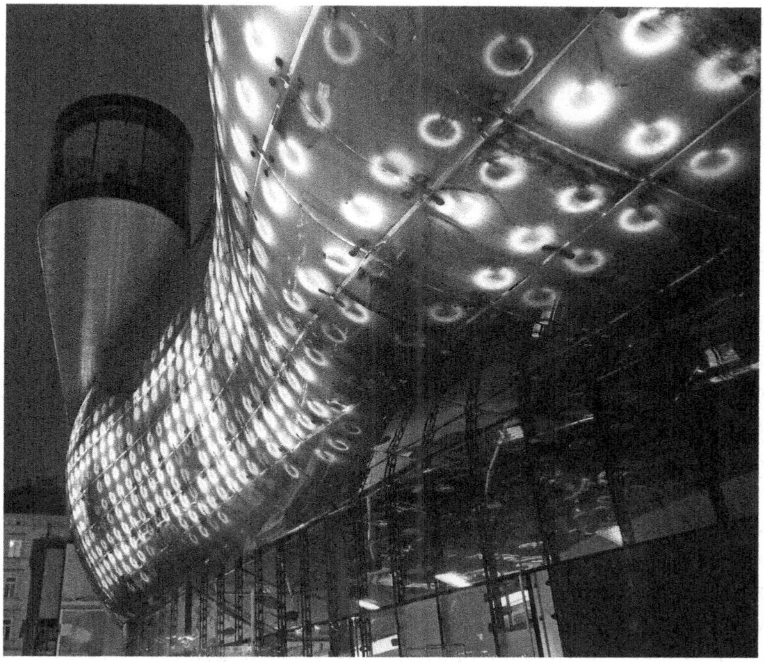

Figure 5.3 Organically formed media facade as a proto-digital skin, Graz Kunsthaus, architects Peter Cook and Colin Fournier (2003).
Source: Author.

between the natural and the digital, an emphasis on code, data flows and networks subsumes nature within the digital. Semiotics leads in a different direction, however. In the previous chapters, I argued that there are limits in applying information theory to biology. Sign systems underpin what we might think of as information flows; not the other way around.

Some scholars attempted to show how computer code fits within a semiotic frame. In an article on computer semiotics, Peter Andersen focuses on the interaction with computer programs as experienced by the user.[30] As in the case of theatre, it is not what happens behind the scenes that carries the signs to the audience, but what is on stage. So, the study of semiotics is relevant to interaction design. He also notes how some styles of programming, such as object-oriented programming, make the semiotics of code more explicit.[31] The concept of classes of variables with properties, arranged in hierarchies and where properties are inherited, as used in object-oriented programming, finds support in the semiotic model.[32] So, computer programming can acknowledge its semiotic legacy.

On the subject of skin, many interaction designers talk about computing as an embodied process, introducing more materially engaged phenomenological understandings of ubiquitous computers and their interactions.[33] In this case, semiotics submits to a more active discourse, though one that I think finds support in the writings of pragmatists such as Peirce, James and Dewey.

Design logics

So far, I have explored biomimesis as a process simply of copying aspects of nature. Biomimetic design also seeks to mimic the *processes* of nature, at least in part, drawing on concepts such as evolution, genetic mutation, competition and fitness, and computational equivalents in terms of iteration, search, constraints, rules and goals.[34] The term 'parametric design' has come to stand for such computationally expensive iterative processes carried out by algorithms that produce organic forms, though as I will show, the approach predates those intriguing, smooth, shattered, bulbous organic shapes to which Perez-Gomez refers: 'Organic shapes, crystalline to folded, simple or complex'.[35]

What is Peirce's contribution to design? Peirce was a logician. Though his work predates computing, he was certainly comfortable with formal logic and mathematics. For Peirce, the theory of signs is a basis for understanding propositions, of the kind used in conventional logic. Logic is a formal language for making assertions and by which inferences can be drawn, and in a manner that is reproducible. That is, anyone else could reach the same conclusions by the same process.

Within the semiotic frame, scholars have described design in terms of Peirce's category of *abductive* reasoning. The architectural theorist Lionel March made the case in a book *The Architecture of Form*.[36] Design does not proceed via a series of inevitable deductions from propositions to a conclusion or solution, but does so evidentially and iteratively. A designer frequently puts forward a hypothesis or a proposition, which is then adjusted, adapted or even abandoned. Furthermore, the evidence sought will develop and change during the process, in the same way that clinicians will adjust their investigation into symptoms considering changing confidence in the hypotheses. I will review Peirce's contribution to design reasoning in more detail in Chapter 7. Rather than Peircean logic, biomimetics generally leads in the direction of parametric modelling.

Parametric design

Biomimetic design draws on techniques of parametric modelling. Parameters are the constants in an equation, set of equations or a computer program.[37] They define and limit what the equation will produce, for example, the shape of a curve. But you can in fact vary these constants to produce new shapes, patterns and objects, like chairs, and biomimetic forms, such as trees, shells and scales. A parameter is different from a variable. According to the the *Oxford English Dictionary* (*OED*) , a parameter is 'a quantity which is fixed (as distinct from the ordinary variables)'. The definition references 'a constant occurring in the equation of a curve or surface, by the variation of which the equation is made to represent a family of such curves or surfaces.' Parameters are associated with curves, but not always. The idea of a family of shapes is important, and buys into the common biological metaphor of families, genera and species. Organisms reproduce to create families with similar traits.

It takes skill to design a chair, but it involves more sophistication to design a procedure or computer program for making many chairs, a family of chairs, with variations. More specifically, it takes expertise to write a program or script that produces a virtual model of a chair that can then be visualized in three dimensions on a computer screen, or even manufactured using a 3D printer. Such an algorithm would provide the user of a CAD system with access to key parameters, such as the height of the seat, the back and the length of the armrests. Varying these parameters produces different designs within a family of designs. I use the example of a chair as it is obvious how the components relate to one another. Organic forms such as shells, trees and scales have similar dependencies within their complicated structures.

With a parametric design, a CAD operator can customize a chair of their own by entering the values they prefer into the computer. The output might then drive a computer numeric controlled (CNC) production line, or a 3D printer, to deliver a chair you can sit on. Much of the skill in parametric design resides in establishing the relationship between parameters. If the chair has no back (it has a height of zero) then the method of fixing the arms would have to change. If the seat is too high, then it may topple over. So, constraints and the relationships between constraints are crucial to the idea of parametric design. The values anyone can code into the parametric system on the computer must be limited therefore. The designer of the program might not let the user decide on the number of legs, and there would be limits on the height of the seat. The various parameters and constraints will also interact. When the designer or user of the system makes the seat project out further then perhaps the arms should extend as well. Some parameters will also conflict. If the chair arms go back too far they will collide with the geometry of the chair back.

People who use CAD systems and building information modelling (BIM)[38] understand the parametric design of chairs well, and CAD and modelling systems provide libraries of parameterized furniture and building elements. A chair is an independent entity in building design, but windows, doors, columns, and skins interrelate in ways that are even more complicated, even just geometrically, especially if those elements repeat throughout the building. So, making the windows 1 centimetre wider has effects that ripple through the whole constraint system that is the final building. In the 1990s, my colleagues and I worked on the logic of constraints in the context of artificial intelligence and expert systems for designers.[39] So-called constraint-based reasoning indeed poses challenges.

Larger entities, such as whole houses or hospitals, compound the complexity of parametric design. Not only are real buildings made up of many geometrical relationships and constraints, but they also involve the selection and arrangement of many parametric components. For example, there are millions of ways of arranging rooms in a building, that is, dividing a rectangle into even just a dozen sub-rectangles.[40] It is not just a problem of enumerating all those possibilities, but of sifting, sorting and selecting the best or most suitable for some purpose or other. Add to the problem of constraints and combinatorics the ill-defined, 'wicked' and random configuration of constraints imposed by environment, context, people, competing stakeholders, social norms and cultural practices.[41] This does not mean design is impossible; just that it is extremely difficult to automate. Automation is at the heart of parametric design.

Biomimetic design

I have already suggested that computation aids biomimetic design. Early advocates of expressionism in architecture worked with biomimetic forms, but were constrained by the drawing and model-making tools available. To create a plausible curved shape in three dimensions such as the petal of a flower requires some intricate geometrical manipulation. A major link between the natural and the digital is that cellular growth follows step-by-step processes: cells divide and build on one another incrementally. Programmers can produce algorithms that operate in a similar manner.

Most people identify the forms of nature as smooth, curvy, repetitive and intricate, contrasting with the products of a straight edge and T-square. Non-uniform rational B-spline curves (NURBS) have developed in importance in architecture and come to represent the organic. Such shapes result from incremental adjustments in parameters looping through computer algorithms. The power of programs, constraints, combinatorics and their limits are well known to anyone who has worked in parametric design. It is no wonder that parametric design flourishes in the production of elegant sweeping building facades and continuous organic roof structures.[42] The biomimetic project is less successful in resolving floor plans, circulation routes and subtle spatial interventions. There are parametric definitions of crowds, swarms and mobs, but as yet nothing that models human sociability and responses to environments in total – the stuff of architecture and landscape architecture. With skins, surfaces and sculptural abstractions the constraints and their interdependencies are more amenable to algorithmic control, unencumbered by issues of use, history, culture, politics and the complexities of human inhabitation.

Despite challenges and limitations, parametric discourse does seem to encourage some designers and theorists and architects to weave it into a totalizing biomimetic philosophy,[43] aided by the complex nature of communication networks. After all, communications systems are networks of so many parameters interacting with one another. Architect Patrik Schumacher has adopted the term 'parametricism' as heralding this approach to architecture. His work with architect Zaha Hadid has resulted in many large, impressive, iconic buildings, such as airports, concert halls, and mixed-use developments, displaying seductively flowing geometries. His theory starts with a simple design trajectory where design is 'facilitated by the attendant development of parametric design tools and scripts that allow the precise formulation and execution of intricate correlations between elements and subsystems'.[44] Such assertions lead rapidly to a highly desirable 'solid new hegemonic paradigm for architecture'.[45] In writing about how such creations adapt to their environments, Schumacher draws directly on a biological analogy: 'The way a single genotype might produce a differentiated population of phenotypes in response to diverse environmental conditions.'[46] Parametricist reasoning draws substantially on genetics and evolutionary theory.

Schumacher thinks 'parametricism' is all pervasive, as it provides parametric knowledge and techniques superordinate to all other approaches. The use of parametric algorithms via computer systems for architecture and engineering has existed since the 1960s.[47] Parametricism is an ancient idea that has its seeds in *idealism*. It echoes some of the mathematician and physicist Roger Penrose's statements about mathematics: 'By some miraculous insight Plato seems to have foreseen ... [that] the actual external world can ultimately be understood only in terms of precise mathematics – which means in terms of Plato's ideal world "accessible via the intellect!"'[48] Idealism is consistent with a belief in the primacy of data as providing access to nature. From a parametricist's viewpoint, the difficulties encountered in parametric design might suggest that design is impossible.[49] But there are other understandings of the design process that appeal to templates, prototypes, narratives, dialogue, play, metaphor, interpretation, embodied interaction and Peirce's concept of abduction. Rather than evolution and growth, some designers also draw on processes of decay in biological systems.

Biomimesis and decay

Contrary to the tenets of harmonious biophilic design, nature disrupts our attempts at order. Left unchecked, nature wreaks vengeance on artifice via invasive plant life with root systems that unseat brickwork, and break through water barriers. Fungus, mould, lichen and other spore-producing organisms (cryptogams) invade architectural cracks, pores and surfaces. As well as beauty, invasive nature invokes a kind of 'disgust' according to architectural scholars Marcos Cruz and Richard Beckett, who say: 'Blotches, speckles and spots of cryptogamic growth evoke visual associations with epidermal disorders, similar to acne or skin sores and rashes',[50] as do changes in the colour of buildings materials and 'excretions and protuberances of growth'.[51] But decay is not entirely deleterious. Cruz and Beckett remind us that nature's slow ravages also evoke a 'bucolic, idyllic vision of nature'.[52] The romantic tradition in architecture valorizes certain biological invasions: ivy-clad walls, moss-covered ruins, patinated stone, weathered and lichen-coated wooden architraves.

Designers can try to resist, or work with, decay. Cruz and Beckett recommend a kind of designed and controlled biocolonization of building materials. After all, such 'green' architecture captures air born pollutants, fixes atmospheric carbon dioxide and nitrogen, is low maintenance, and can even generate energy. Computer modelling and digital printing aids their practical research into the design of 'bioreceptive' materials and surfaces to foster a kind of 'biodigital materiality'. They are working on intricate surface geometries and 'bioscaffolds' that encourage certain configurations of growth, and channelize moisture to encourage cryptogamic growth. Cruz and Beckett think of crusty and furrowed tree bark as a fitting model of what they are trying to achieve:

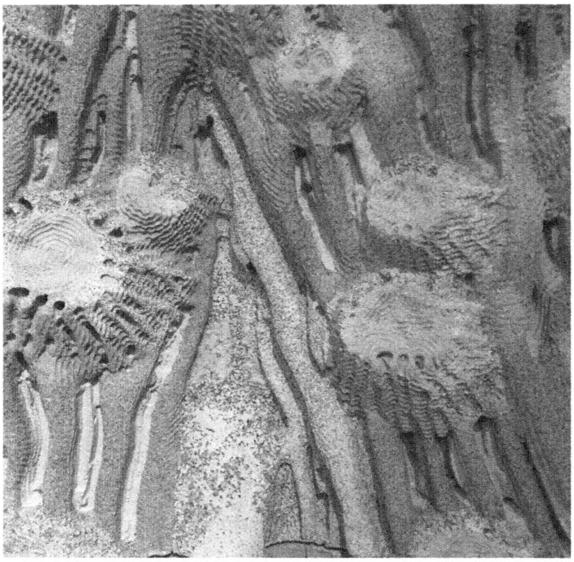

Figure 5.4 Bioreceptive Calcareous Composite Wall by Zhili Wang, Xinhe Lin, Yuxin Jiang, and Qingyue Zeng at the BiotA Lab, Bartlett UCL (Prof. Marcos Cruz and Richard Beckett). This is a multilayered cast of a bioreceptive prototype with use of different particle sizes to enhance a selective water-retention system in a wall. Photo credit: Marcos Cruz.

> Areas of shadow, areas of protection, crevices that trap dust and nutrients and water channels are all typological variables that occur on tree barks and provide very specific conditions at the material surface which allow for or restrict growth.[53]

Many of these innovations are outside of the usual appeal of large-scale structures assuming elegant, flowing organic forms. A browse through images of biomaterial processes reveals algae-stained glass surfaces, fractured membranes, globules, stains and slime. Grotesque bio-architectures also emerge in the work of architectural researcher Neri Oxman. In the Silk Pavilion project her team devised technologies to encourage silkworms to deposit threads across frame structures in response to controlled light conditions.[54] Such innovations flirt at the edges of the elegant and the grotesque.

Cruz and Beckett's approach to recruiting microorganisms that colonize building surfaces is in the company of ideas about dynamic 'living skins'. Emma Flynn provides an account of experimental building technologies for mediating between internal and external environments.[55] One such system involves walls clad in photobioreactors containing algae that capture carbon from the atmosphere, and combined with sunlight to collect heat, provide dynamic shading and produce biofuel. Other innovations include self-healing wall coatings made of synthetic photocells that form a crystalline microstructure like limestone when exposed to carbon dioxide, offering the possibility of 'healing' stress fractures in buildings – a process analogous to the operations of organic 'scar tissue'.[56] In an extensive review of organic, 'vibrant' architecture, Rachel Armstrong describes a similar process in 'bioconcrete':

> The hardy organisms mixed into the cement are activated when tiny cracks in the concrete let in water and produce a calcified sealant that prevents further progression of the micro-fractures.[57]

She also describes a 'membrane growth and repair process'[58] where osmotic forces rupture the cell membranes 'that form around the copper II sulphate crystal and immediately heal as the salt solutions come into contact with each other'.[59] There is a place for concepts of decay and healing in any survey connecting nature with the digital in architecture and landscape architecture.

The limits of biomimesis

Does good design always mimic the things of nature? There is as much pictorial evidence for sustained positive responses to the artificial as there is from the biomimetic. Le Corbusier's tubular steel chair is much admired, though it makes scant claims for connection with nature. Nor does Michael Graves's chrome whistling kettle, or an iPad. Comfort, novelty and function come into play in good design, but every successful design does not need to be aligned with nature, or biophilic design. In fact, much good design works through contrast and difference, and can even enhance a sense of the natural, not by imitating it, but by presenting as its opposite. Perhaps it is the stark contrast provided by an alien, artificial and 'unnatural' object in the landscape that brings the natural into relief – draws our attention to that context as 'natural', and defines the natural for that place and time. Sculpture often does this, but so can buildings, bridges and motorways. This was one of the messages of Sylvia Crowe in her classic book *The Landscape of Power*, in which she encouraged motorists and walkers to appreciate those giant cooling towers and gasometers (that are now disappearing) in the landscape.[60]

In Scotland, architects from Robert Matthew and Stirrat Johnson-Marshall (RMJM), Arups, and Butterley Engineering designed the Falkirk Wheel that raises pleasure vessels from one canal system to another by means of a rotating elevator. A few kilometres away on the same canal there is a 30-metre high steel sculpture of two horses' (kelpies') heads (by sculptor Andy Scott), which is part of a landscape regeneration scheme. Concepts of harmony are not so prominent here. Though their forms are derived from nature, these elements in the landscape operate in a way more akin to the revealing of difference. Martin Heidegger advanced the view that things in the landscape reveal something important about their context: 'The bridge gathers the earth as landscape around the stream.'[61] He was talking about stone bridges. But recognizing Heidegger's overtly romantic references, philosopher of technology (and post-Heideggerian) Don

Ihde asserts that this capacity of things to reveal place, nature, ecology and environment applies to objects of steel as much as to stone, and even to nuclear power stations, especially if we take politics into account.[62]

Interpretation comes into play. As for any sign, harmony, wholeness and unity may be as much in the eye of the beholder, or the camera, as they are properties of the place. For example, any of the pictures in Thomas's *Biophilic Design* could be photographed under lighting conditions and in such a way as to look less harmonious, and they could be captioned or described to make the case either way as to their visual, structural or natural integrity.[63]

In many respects, it is of little consequence that ideas of the natural and the digital are conflated. But I do not think this conflation is a satisfactory response to their interaction. That people think of the natural and the artificial as categories in the first place means that, rightly or wrongly, society attaches significance to the terms, and the distinctions they entail, that is, finds them useful in some way. Expressions such as 'back to nature', 'natural attitude', 'it's only natural', 'natural language', and 'the world of nature', have use value, as do terms like 'digital age', 'digital media' and 'going digital'. It is a basic tenet of semiotics and its related movement in structuralism to take such oppositions seriously. These oppositions at least signal power relationships that cannot be glossed over, trivialized or ignored.

I hope I have shown that data and algorithms offer a potent ensemble for making places through concepts of biomimesis and parametric design. Some scholars and practitioners elevate parametric design to the status of a world view that aligns the complexity of communication networks with the intricacies of organic systems, the complex myriad of communicating and interacting cells and organisms that make up life. Parametricist reasoning invokes genetics and evolution to bolster its credibility, attempting to draw together data and nature on the path to biomimesis, following the autopoietic, self-making properties of nature. In this chapter, I reviewed critically the human propensity to be receptive to such claims, which in turn reflect on our attitudes to place and place making.

CHAPTER 6
THE BOOK OF STONES

Communicative structures and processes pervade life systems. Biosemiotics promotes the interconnection of all living things. But are rocks, mountains, magma, and the material worlds of earth-bound and extraterrestrial geology engaged in comparable communicative processes?

Though many architects and engineers are interested in biomimesis, buildings and bridges have more in common with geology than biology. I take this conclusion from a series of interesting articles in a special issue of *Architecture Research Quarterly* (*Arq*) on architecture and biotechnology. More accurately, the skeletons, hardened excreta, dead tissue and shells provide the structural support for organic life. They outlast the living organisms that produce them and provide ready analogues for human-made structures. Such rigid materials also associate more readily than soft, motile tissue with the lifespan of a building. Some organic material eventually breaks down and turns back into nonliving carbon deposits, limestone, flint, marble and other substances that end up in buildings. In any case, the constructions that have survived from antiquity tend to be made of geological material, that is, hard stone.

Amid an enthusiasm for an architecture that mimics biological life, the scholar of experimental anatomy Jamie Davies outlines several points at which architecture and biology part company.[1] Whereas buildings are constructed following plans in a top–down manner, organisms develop piecemeal, responding to local conditions from the bottom up. Organisms also rely on feedback from their environment via small interactions that then propagate to the formation of the whole organism. That much is obvious. But the fact that buildings are not alive turns out to be one of architecture's major assets. After all, because buildings are nonliving, the functioning of a building can be halted temporarily for maintenance and improvement. Unlike zygotes and embryos buildings do not have to start functioning until they are completed: 'Human-designed buildings and other machines have to … function only once they are completed: furthermore, function can be suspended when maintenance and alteration have to take place.'[2] Davies adds, 'Developing bodies, on the other hand, have to be viable throughout.'[3]

However, the idea of buildings that live and breathe, adapt to their surroundings and either incorporate living material or integrate synthetic structures and materials on organic principles has some appeal. Some architects even propose nanobuildings 'that will be "grown" from the subatomic level' and start their life in a vat.[4] But living things die, can be killed off, and are extremely sensitive to environmental conditions; 'while the idea of "growing your own building" may be attractive, a truly living building would have the disadvantage that individual living things are easily killed.'[5] Davies invokes P.B.

Shelley's poem *Ozymandias*, where all that remained of the mighty Pharaoh was a stone statue in ruins on the desert floor. Davies remarks, 'as fossilised shells silently state with a different kind of eloquence, stone artefacts long outlast their organic builders'.[6] In this chapter, I will champion the cause for geology as a major component of communicative commerce with and through nature, as well as artificial constructions such as buildings, engineering works, digital devices and automata.

Mountains that speak

On the subject of geology, permit me to relate my own commerce with a mountain. Munros are mountains in Scotland whose tops are at least 3,000 feet (914 metres) above sea level. The only time I scaled a Munro on my own was in late summer 2003 when I traversed a horseshoe of ridges to reach the peak of the shallow cone that is Ben Cruachan (1,126 metres). On the peak, I was in the clouds, and the route looked the same in all directions. The clouds cleared for a moment and I returned the way I came, rather than continue around the horseshoe to complete the circuit. Unable to read the terrain I started to follow a ridge that veered into a different valley and away from my starting point. When the terrain became less familiar and steeper, I decided to clamber to the ridge again. I calculated that I would not be back to my car before dark. I called my hotel on my mobile phone to tell the Mountain Rescue Service to search for me if I was not back by 7.00 pm. You can be more certain of a cellular connection when on a ridge than when lower down in the radio shadow of a mountain. Then the cloud cleared, and I saw below the distinctive water of the reservoir and the dam wall over which Ben Cruachan and its smaller compatriots preside. The half ring of mountains surrounds the reservoir, and I had my bearings. Some hours later I reached the safety of my car, only scratched from walking in the dark through brambles along the flatter reaches of the valley.

That reservoir supplies water for the Ben Cruachan hydroelectric power station completed in 1965, with its turbine hall secreted deep beneath the mountain. To commemorate the fiftieth anniversary of this feat of engineering, colleague and writer Maria Fusco was commissioned by Artangel and BBC Radio 4 to create a performance piece in the machine hall. She called it 'Master Rock', described as: 'Fusco's repertoire for a mountain fuses sound from Cruachan power station with three spoken voices: an Irish tunnel tiger, a forgotten artist, and the voice of the 450 million year-old granite itself'.[7] I am at ease with the thought that the mountain spoke to the writer, as it did to me, the neophyte hill walker, particularly through topography, atmospheric conditions, warnings and other signs. I also understand that the mountain's prospect both aided and blocked mobile electronic communication. Fusco's proposition that her work mobilized 'the voice of the 450 million year-old granite itself' provides an interesting challenge for semiotics. In her performance, the voice of the mountain was mediated by a vocal artist, with live manipulation by a sound composer to suggest ascent and descent, cracks as the

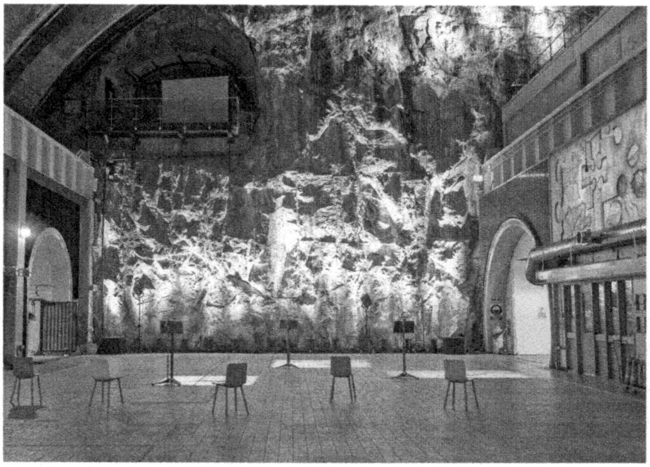

Figure 6.1 'Master Rock' site-specific performance 15 October 2015. Courtesy of Artangel and Maria Fusco. Photograph by Robert Ormerod. Used with permission.

earth moves, electrical discharges and ethereal resonance suggestive of a cave as much as solid granite. The geology–human semiotic relationship is easy enough to appropriate, particularly when mediated by interpretation from a human performer, and with sensor, amplification and broadcast technologies – but what of intergeological communication? Does it make sense to think of semiotic relationships among geological structures and processes, independently of human mediation?

Geosemiotics

Charles Sanders Peirce's day job was that of a geologist in the US Coast Survey. Peirce did not explicitly relate his theories of signs to his practices as a geologist, though his knowledge of geology influenced his philosophy. As part of their professional skillsets, geologists study forms, samples, sediments and fossils, and read the geological signs to identify processes, invisible substrates and structures. In an influential address to the Annual Meeting of the Geological Society of America in 1998, the geologist Victor R. Baker outlined the communicative functions of geology, with full reference to and expanding on Peirce's philosophy. Baker laid out the basis of *geosemiotics*, advancing the strong Peircean thesis towards 'a semiotic that is continuous from the natural world to the thought processes of geological investigators'.[8] After all, geological practices are amenable to description in terms of Peirce's classification of signs. For example, photographs, drawings and maps constitute iconic signs. The character and appearance of a rock sample will divulge some process of which it is a causally linked indexical sign. So, a thinly layered (foliated) rock sample is a sign for (in geological terms) 'a stratigraphic sequence, in which its indexical character relates to temporal succession'.[9] Geological symbols come to light via human interpretation: 'The words and mathematical expressions of science all have symbolic character, such that much of the applicable logic has concern for the manipulation of symbols'.[10] Geological big data is a symbolic system after all. The communicative relationships between geological systems and human geologists are easy to establish. But how do we explain communicative relationships within geological systems?

Thomas Sebeok could explain biological signs in terms of communicative mechanisms among plants and animals independent of human interpretation, including the communication of symbols and names. It is tempting to think that communication among nonliving geological elements is more limited than among living organisms. Baker's geosemiotics advances the post-human idea that all of nature is made up of a web of intercommunicating sign systems independently of human control, but in which human beings participate from time to time. For Baker, we humans are part of 'a semiotic web, according to which things and objects interweave to make up the fabric of experience'.[11] Supported by Peirce's semiological project, and Dewey's work on experience, it is as if experience and thinking reside in the larger environment, rather than something overlaid onto it by human beings. For Baker, the ways geologists reason 'constitute sign relationships that extend continuously from the physical world of what is observed to the mental world that is generally associated with observers'.[12] What applies to expert geologists surely applies also to the rest of us.[13] By this reading, we human interpreters are part of a communicative network that includes not only the exchange of signs within the biological world, but the world of geology and communication within geological systems as well, which are commonly described as 'nonliving'.

I referred to 'the book of nature' in Chapter 3. In 1863, David Thomas Ansted (1814–80) put the spotlight on geology with his book *The Great Stone Book of Nature*. Anticipating aspects of semiotics, Ansted invoked metaphors of communication to account for geological processes. Movements in one part of the earth's crust are communicated to another some distance away, with obvious presentations as earthquakes and geothermal eruptions. Not least in this commerce are the effects of water, at least near the surface of the earth 'as it circulates through the earth, as it passes from the sea to the sky, and as it returns in refreshing showers from the sky to the earth, and so back again to the ocean'.[14] He affirmed that this circulation serves as the 'means of communication', providing 'the connecting link between the power and the conveyer of power'.[15]

More recently, the sociologist Bronislaw Szerszynski has enlisted Ansted's book in support of the semiotic functioning of nature, and its implications. Drawing on Ansted's book, Szerszynski alludes to the page-like character of layers of geological sedimentation and the identification of signs such as material deposits, folds, metamorphoses, fault lines and fossils. The geologist rifles through the layers of time as if turning the pages of a book. The metaphor goes further. A book in your library is made up semiotically of 'sentences, paragraphs, sections, chapters and parts'.[16] The book of geology is also 'divided into nesting material wholes – stages, series, systems, erathems, eonothems'.[17] But the material structure of the earth corresponds with its semiotic structure, 'since each unit of time-rock can be read to reveal corresponding nesting periods of rock-time – ages, epochs, periods, eras, aeons, each divided by moments of dramatic change in the Earth system'.[18] The earth's geology is available for the geologist to be read just as chapters and sections of a book are available to all readers. Geology tells a story, with highlights, moments of calm and cataclysm, and with several interweaving narrative threads – and the story is not yet over. Of course, contemporary geology has moved on since Ansted's *Great Stone Book*. Szerszynski shows how contemporary geology advances on the book metaphor, as we think of the earth as a system, with diverse surface features resulting from 'slow, invisible unifying forces such as sedimentation, volcanism and tectonics'.[19]

Human influence

Though geology is infused with communicative structures, as human beings we are bound to show greater interest in our own communication with geology. For Szerszynski, the earth is subject to 'the centrality of the geological "gaze" in the field, the linear but contingent deep history, the constant move from surface differences to deep unities'.[20] But we do not just read the stone book of nature; we are characters in its slowly unfolding drama. Stretching the book metaphor further, we are also its writers. We are entering a stage in geological history where the reader writes the book.

There is a practical urgency to the relationship between reader and nature. The relationship comes to the fore in the idea of the Anthropocene, designated as the most recent geological period. It is the current epoch in which human agency leaves its mark on the geology of the planet, a process comparable to an ice age, or other major identifiable, cataclysmically defined epoch.[21] Technically, we are now at the Holocene period, which began 11,500 years ago as the glaciers began their retreat. The Anthropocene supposedly comes after the Holocene, or is the most recent stage in that epoch. The term Anthropocene is still controversial and it does not yet appear as an official geological designation. Not least, geological periods are retrospective, and of long duration, whereas the presumed Anthropocene is a relatively new 'epoch', and insufficient time has passed to assess its significance as a geological category. Nonetheless, the Anthropocene features as an important concept in powerful critiques of the human's relationship with nature. An article in the journal *Nature* foregrounds challenges in geological definitions of the Anthropocene stating: 'The impacts of human activity will probably be observable in the geological stratigraphic record for millions of years into the future, which suggests that a new epoch has begun.'[22]

The actor in this geological story is of course the *anthropos*, the man/woman, the human animal, or homo sapiens. The anthropos is the agent who devises the geological classifications in the first place, but also brings about global transformations, changing the earth's geology. As the technologized human, the anthropos impacts geology by introducing agriculture, mining the earth, damming rivers, altering coasts and changing climate. We are at a stage in the history of the world where the anthropos both reads and writes the very long stone book.

The Anthropocene

Returning to Szerszynski's lessons from *The Great Stone Book of Nature*,[23] semiotics provides a useful framework for making sense of the Anthropocene. He argues that 'the truth of the Anthropocene is less about what humanity is doing, than the *traces* that humanity will leave behind',[24] that is, the signs and signals that get inscribed in 'the great book', in the rocks of the earth. He invokes the semiotic language used by Zalasiewicz et al, which first articulated the case for the Anthropocene.[25]

Zalasiewicz et al identified the 'lithostratigraphic signal' resulting from changes in the courses of rivers, coastlines and water flows engineered by human settlement. In technical terms, there is the 'chemostratigraphic signal' indicating our alterations to the chemical composition of air, land and sea, including the results of pollution, the increase in reactive nitrogen in agriculture and the production of novel compounds due to nuclear fission and other industrial processes. The 'biostratigraphic signal' emerges from species depletion, agricultural monocultures, and over-fishing. Then there are changes to the 'sequence stratigraphic signal' produced by large sea-level rises.

Zalasiewicz et al did not reference semiotics, but as if to reinforce the book metaphor, they explained the reference points geologists place at exemplary cliffs, cuttings or outcrops in key sites on the earth's surface. These markers are like bookmarks. There are around 100 such sites known as 'Global Stratotype Sections', Global Boundary Stratotype Section and Points (GBSSPs) or 'golden spikes' located at agreed-upon points on vertical rock faces. These durable and appropriately tagged metal spikes indicate the sharp demarcation between geological periods as 'time planes'. They mark 'an elapsed, distinctive, and correlatable geological event rather than an arbitrary or "abstract"

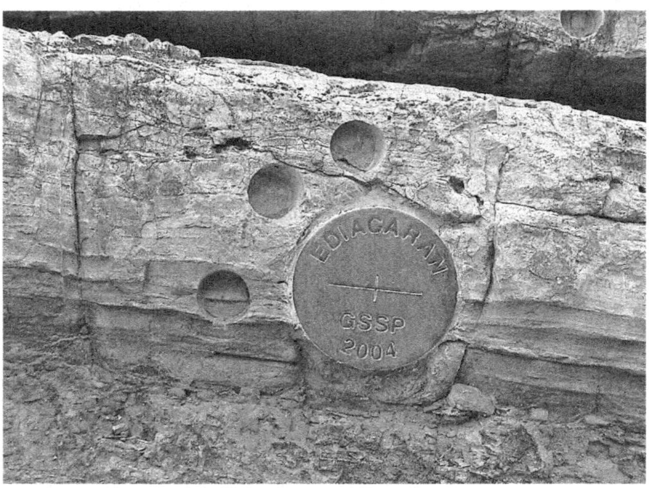

Figure 6.2 The 'golden spike' marking the Global Boundary Stratotype Section and Point (GBSSP) at the base of the Ediacaran Period 16 August 2008.
Source: Bahudhara via Wikimedia commons.

numerical age'.[26] So for example there is a GBSSP for the Kimmeridgian stage of the Upper Jurassic period, in Flodigarry on the Isle of Skye in Scotland at 57.6000°N 6.2000°W. For Zalasiewicz et al the question arises as to where such golden spikes would be located to bookmark the start of the Anthropocene period, and at what stratum. The location of such markers depends on the time period demarcating the commencement of the Athropocene. Was it the start of deforestation, the Industrial Revolution, or the first indications of climate change? The authors also reference the globally identifiable level provided by 'the global spread of radioactive isotopes created by the atomic bomb tests of the 1960s', and the natural event of 'the eruption of Mount Tambora in April 1815'. The latter was not man-made, but the deposition of air-borne sulphate left 'a distinct global signal in the dendrochronological record'[27] and serves as a robust surrogate time marker for the start of the industrial era. In terms of geosemiotics, the volcanic deposition would serve as a *symbol* for the start of the Anthropocene.

Szerszynski thinks of signs as bookmarks: 'These spikes are thus like permanent bookmarks in the stone book of nature, marking the boundary between its parts, chapters and sections.'[28] He favours the view that the Anthropocene 'in all its geohistorical specificity really starts when humans become aware of their role in shaping climate, and this awareness shapes their active relationship with the environment',[29] with major interventions such as genetic modification and geoengineering still to come.

The Anthropocene brings into relief the idea that we influence the course of the book of nature's narrative, and not just by planting bookmarks. Szerszynski puts it thus: 'As the anthropos turns from reading to writing the stone book of nature, this is a "being written" that seems to disrupt the order and meaning of all the other pages of that "written being."'[30] The anthropos is therefore multiple, underdetermined, technologized and part of the geological fabric through which its being is described and recognized. It is worth capturing something of the poetics of his geosemiotic discourse by providing a fuller quotation:

> What we as humans put down in the stone book is the disruption of other layers, a rifling through the pages, as we drill, mine and extract. We are volcanic, creating extrusive and intrusive formations that break the logic of superposition and burst the relation between space and time in the stone book. Just as magma fills fissures and then cools to create 'dikes' – thin sheets of igneous rock that lie discordantly across existing strata – we create pages at strange angles, generating a 'Rubik's book' that would need to be read through in all directions simultaneously. The Anthropos will thus 'lie' in the strata in a different sense, in a different plane, not 'true' – as a perjurer, disrupting the semiotic logic of geology as much as its materiality.[31]

Having identified the Anthropocene, we must acknowledge that there are several orientations to it. There are the 'deniers' of course, who refute any significant human influence on earth systems, and are generally out of the discussion. But among the acceptors there are several categories of response. Following the stone book metaphor, we could think of these as 'authors' of the Anthropocene. I have distilled the categories that follow from those devised by Anthropocene scholars such as Thomas F Thornton and Yadvinder Malhi, Charles Travis and Poul Holm, Michel Serres, Lesley Head, and Szerszynski. In my own version of these Anthropocene respondents I identify the interrelated categories of the modernists, pragmatists, melancholics, earth children, and subversives.

Modernists

Geographers Thomas Thornton and Yadvinder Malhi identify several key modernist responses to Anthroprocene challenges.[32] There are the predictors, or prophets, who attempt to sustain a sense of alarm and predict disaster. Some foreground progress as inevitable and costly, but think it will ultimately save us. Others believe, either blindly or with knowledge, that we can fix the planet's ecosystems with technology. This is the world of the

> Technofix Optimist, which emphasises climate change and other Earth System problems as solvable through human ingenuity and technology, whether through solutions such as cleaner energy supplies, more efficient urbanisation, or global geoengineering.[33]

Some modernists advocate for new ways of thinking and acting. This group advocates for a 'fresh vision of re-enchantment emerging from the Anthropocene in which humans reconnect with Earth systems'.[34] Thornton and Malhi point to the "Ecomodernist Manifesto" in which we read:

> A good Anthropocene demands that humans use their growing social, economic, and technological powers to make life better for people, stabilize the climate, and protect the natural world.[35]

As further candidates for the modernist *anthropos* of the Anthropocene, Szerszynski offers the maker (*homo faber*), the consumer (*homo consumens*) and the helmsman (*homo gubernans*). Gubernans is also the governor, manager or pilot in the sense we find the terms used in sailing. The idea of the helmsman suggests an agent engaged with the materials of that which she wishes to change, as is the case with someone who steers a boat through currents, or as a potter manipulates clay, the outcome of which depends on interaction between the properties of the clay and the deliberate actions of the potter.[36] This complicity between the maker and the tools suggests that we abandon the view that humans can ever exercise complete control over nature.

Szerszynski draws on the philosopher and provocateur Michel Serres, who advocates for the helmsman or pilot who follows a designated route but tilts the rudder 'depending on the direction and force of the swell'.[37] Human will and its obstacles engage in a to-and-fro exchange 'in a series of looped interactions'.[38] As reader and writer of this book of stones humans steer the world through geological time.

Pragmatists

Following Serres, the *anthropos* can operate as a pragmatist. Geographer Lesley Head thinks of such engaged people as 'the well-off citizens of the Modern world who, having contributed so much to the problems, have to try and remake ourselves and our worlds'.[39] Pragmatists are not those who hope for something better, but get on with the business of change. Drawing on several scholars who advance the theme of *hope*, Head concludes that we need to move beyond the idea of hope as something you feel, to something you do. Apart from fuelling major transformations, this emphasis on the practical demonstration of hope helps the individual make sense of those small gestures that may so far have little impact on improving the environment but circulate and signal hope. We might think of dividing up our recyclables, cutting down on air miles, or abandoning the car and catching a bus to work.

Sensible governance and management build on ideas about stability, predictability and balance,[40] and of course, these motivate building and architecture. But Head thinks we may be moving towards a future without these stabilities, where transience and mobility become the norm:

> On the one hand the hyper-mobilities of late modernity are a key contributor to greenhouse gas emissions; on the other hand sea level rise and other changes mean that whole societies will need to be on the move.[41]

So, the challenge is to adapt to the loss of spatial stabilities (perhaps tested with refugee migrations across the world). She labels such pragmatists 'Anthropoceneans', who draw on the past and heritage, but do not want to return to the past. In breaking with past methods, pragmatic Anthropoceneans believe there is scope for judicious and well-researched extreme measures.

For Head, it 'is now clear that the metaphor of treading lightly on the earth does not actually help operationalise turning around this Titanic'.[42] We need to hit the ground with a thump. We are ontologically connected to the earth, but also in the practical sense of managing resources, agriculture and water supply. She was writing from the Australian context in which indigenous approaches to the land are pertinent politically. We occupy 'multifunctional landscapes with many overlapping land tenures and understandings'.[43] Not everything we might do is appealing or family friendly: 'We should not only focus on the gentle – vigilance, killing and culling are part of the package. Much environmental work is labour intensive, whether killing invasive weeds in the savanna or juggling household activities to reduce car use'.[44] As well as developing renewed understandings of the past, some of us will have to adjust what we do with our time: 'Provision of food and water will probably take more hours of the day, leaving less time for commerce, formal education, cultural pursuits.'[45] Some of us will obviously have to share what we produce, be prepared to draw on the generosity of strangers, and share jobs.

Head was writing before recent oversupplies of oil, but it is evident that we moderns are not very good at dealing with abundance, with investing for leaner times and with fluctuations in the oversupply of water and agricultural products. Head recognizes that audiences need to be engaged beyond what is offered by 'environmentalism'.[46] We do not need to join organizations. Sometimes such organizations act as barriers to engagement.[47] Anthropoceneans do not indulge utopian or idealistic narratives of what needs to be done, but are sceptical pragmatists. Such is the complex response of the pragmatic Anthropocenean in reading and writing the book of nature.

Melancholics

Professionals, campaigners and activists have a role as emotional labourers.[48] Head says, 'Grief and other painful emotions – fear, anxiety, trauma – will be our companion on this journey – they are not something we can deal with and move on from.'[49] There is sadness in the Anthropocene, especially in the face of those who deny the severity of human influence on climate.[50] Her focus on the emotion of grief leads her to say, 'the first step is to acknowledge this companion, grief'.[51] She adds solemnly, 'If part of what we are grieving for, and what we must farewell, is our modern selves, it follows that a necessary intellectual and practical task is to imagine new kinds of selves.'[52] She confirms the obvious point that we human beings are not after all 'the centre of things – the earth does not actually care whether we survive or not'.[53] If the Anthropocene is a period of loss and grief,[54] then what have we lost? Who are 'we'? According to Head, 'The divide between anthropos and other (usually thought of as humans and nature) is one of the many connected dualisms that must be undone and rethought.'[55] To assume it is all about us is part of the problem.[56] Human beings reflect, articulate and worry through the Anthropocene, but there is a clear case for solidarity and empathy with the rest of existence.[57]

Grief provides a common anchor point for communities and cultures. As far as we know, other species also display what we think of as grief. She says it is obvious in elephants and crows, for example.[58] Head does not express it like this, but there is a sense in which 'the whole creation groans and suffers' for redemption.[59] The feeling of loss in the Anthropocene extends beyond the human species, and by this reading is a unifying communicative strand in network nature.

This focus on grief attempts to unseat the Enlightenment definitions of rationality that partition and then exclude those facets of being that we commonly associate with the emotions. Studies in human geography, environment and the arts are undergoing an 'affective turn' – a reintegration of affect, emotion and mood into considerations of human being, place, environment, nature and technology – and in my own studies into digital technologies.[60] In the face of the incidence of catastrophic bushfires in Australia, Head is concerned in particular with 'the emotional dimensions of climate change'.[61]

The losses she refers to are for loved places, particularly homes and homelands, due to environmental change. Then there is non-human loss, of species and the relationships between them. Extreme events such as earthquakes and bushfires cause intense, sudden and high-profile loss and trauma.

But the most significant and enduring loss is of modernity itself, and its claims to rationality. In the company of other commentators, Head says she is 'grieving for modernity, for a future that was always foggy but was presumed to contain the seeds of positive possibility'.[62] There is a link here with the cultural theorist Jonathan Flatley's identification of the current period of melancholy, lament over losing the innocence of modernity – the naive belief in progress.[63] In this loss there is the inevitable denial, acting as if nothing is changing: a position brought into prominence through climate-change deniers.

For Head the focus on grief shifts the problem of climate change into the realms of therapeutic discourse.[64] Though she recognizes that the condition in the Anthropocene is less about finality, Head commends the need to come to terms with loss – to get past the stage of denial 'to find ways to carry our grief into hopeful environmental engagements'.[65] As if engaging in a kind of collective therapy, the anthropos needs to probe 'exactly what we are grieving for'.[66] She says that simply looking to the past is not the answer: 'Against such an ideal, the present can hardly be understood in any other terms than loss.'[67] Head advocates a response to loss that moves beyond the linearity of time. She draws on the writing of my colleague Michelle Bastian and others to affirm the importance of multiple temporalities.[68]

It is worth turning again to Serres, who constructs one of the most melancholic characterizations of the anthropos as the *defiler*. Serres sees man as a parasite in relation to environment, but also as one who marks his territory by leaving behind bodily secretions.[69] Serres observes that many people 'mark and dirty, in a kind of defecation, the objects that belong to them in order to keep them, or other objects in order to make them their own'.[70] He says it is like spitting on your own salad so no one else will dare touch it. Marking territory is indisputably a semiotic act, where leaving such traces constitutes signs of possession. This is not bookmarking the pages of nature but staining, tearing and scribbling on them. He says of such deliberative and hegemonic human acts that 'the dirtying of the world imprints onto it the mark of humanity, or of its dominators, the filthy seal of their conquest and appropriation'.[71] Such suspicion speaks vividly of the anthropos as melancholic. After all, for human beings and their habitats there is futility in such defilement. It is a situation where, in the face of having what we want we deliberately contaminate and sully it. Melancholy is a sense of misery and dissatisfaction in the face even of success. As Sigmund Freud would say, it is loss, and the anxiety of loss, even when we have what we seem to want.[72]

As I have explored elsewhere melancholy connects spatially with the earth via the horizon.[73] The ground rises into prominence when we are aware of its expanse and scope, viewed as a clear horizon. The horizon carries connotations of travel, yearning and loss.[74] I have also referenced Lars Von Trier's disturbing film *Melancholia*, which is about the end of the world as experienced by guests at a wedding party in a remote and opulent resort hotel. The earth is impacted by another planet, visible as an earth-like orb whose horizon is clearly visible during its slow and inexorable approach. This is a fatalistic story that things might end irrespective of what we do to our planet. The loss of earth, life, planet presents as a surrogate for homesickness, the loss of connection with our anthropos origins. Anthropos is after all the gnostic term for the first human, Adam, born from the dust of the ground and into innocence – in tune with nature, but then ejected from the Garden.

Children of the earth

My fourth category of response to the Antropocene engages with the very earth under question, and upon which the anthropos bears its influence. It is also the earth that gives birth to the anthropos. I hope the reader will indulge this foray into the heart of myth and conjecture. The sources I draw on suggest that such stories access the substance of our being in the world. In any event, they connect aspects of culture, ancient and modern, that in turn inform our responses to nature and environment.[75]

I have referred already to the being who is in tune with nature, an attunement that extends to the material earth on which we dwell.[76] The tuning metaphor surfaces in geosemiotics. According to Baker we human beings, and the scientists among us, can see and reason about regularities, relationships and connections in the universe 'because the human mind is instinctively attuned to nature'.[77] Such an attuned creature is a particular kind of anthropos, namely an *autochthon,* someone born of the earth. Without drawing explicitly on scientific method or semiotic theory, environmentalists such as David Abram point to the human's innate affinity with the earth, and the earthiness of existence:

> We can sense the world around us only because we are entirely a part of this world, because – by virtue of our own carnal density and dynamism – we are wholly embedded in the depths of the earthly sensuous.[78]

He refers to our ability to experience 'the tangible textures, sounds, and shapes of the biosphere because we are tangible, resonant, audible shapes in our own right'.[79] He continues with reference to earth and soil: 'We are born of these very waters, this very air, this loam soil, this sunlight.'[80] Sometimes our objectifying language denies this earthiness, as when we speak in terms of earth's resources as something to be measured and exploited. But the connection with the earth is real and represents a 'consanguinity', that is, a shared lineage, to be honoured: 'We are neither pure spirits nor pure minds, but are sensitive and sentient *bodies* able to be seen, heard, tasted, and touched by the beings around us.'[81] Consanguinity is an interesting term in this context. As well as implying blood relations and affinity, it relates to the sanguine personality trait, 'a courageous, hopeful, and amorous disposition' according to the *OED*, and is commonly set in opposition to the melancholic.

There is little in our daily experience that does not at some stage rely on the ground for support, that succumbs to its insistent pull, or that emerges from it, only to return.[82] We know now that the world of plants and animals exists as a thin veneer on this massive block of ground, the inner core of which is earth under pressure, occasionally surfacing as molten rock. Plant life also issues from the ground. It is no wonder that the ground features in myth and metaphor as a source of living things.[83] Earth, language and origins coalesce in the mythic imagination. After all, Adam was formed of the 'dust from the ground',[84] and the philosopher Socrates asked how parents produced offspring in the early days. His counsellor said to him,

> Quite simply, Socrates, they didn't: there was no such thing at that time as parental procreation. It was the earth-born race, whose existence once upon a time we hear of in our stories, which was born: that was the time when they began to rise up again out of the earth.[85]

According to the *OED*, an autochthon is a person traditionally born directly from the earth, or whose ancestry can be traced back to the earth. The term is also used for an 'indigenous person', an earliest known inhabitant, that is, someone who belongs in the place, speaks its language and is in tune with its communicative structures.

The ground as origin competes with the more obvious source of living things – animal organic reproduction. We could excuse someone unaware of processes of procreation and evolutionary biology for thinking that living beings emerge from the ground, as suggested by Socrates's counsellor. One of the stories disingenuous parents tell children about the origin of babies is that they appear beneath cabbage patches. A living thing not produced by parents has a claim on the ground as source and origin, even if manufactured by a human agent.

We can update myths of autochthonous origins to romantic and contemporary fiction. Frankenstein created his synthetic human from body parts exhumed from graves.[86] Authors of cyborg fiction capitalize on the link between ground and artificial lives. In the case of Coppelia the living doll there is the connection with alchemy and the philosopher's stone, or primary matter. As well as falling to earth, there are those occasions where something falls into the earth and then returns to the surface. In the original novel of *Pinocchio*, the puppet enters hell and undergoes rebirth.[87] Entering the ground and emerging from it features as a common narrative trope in the context of puppets, mechanical dolls and automata.[88] Something similar applies to robots and human–machine hybrids. 'Cyborg replication is uncoupled from organic reproduction', wrote Donna Haraway. It is obvious that clones, machine intelligence, androids and autonomous half-human robots, in actuality or in prospect, are produced by means independent of sexual reproduction.

According to the structuralist anthropologist Claude Lévi-Strauss, creatures born of parents are 'born from different', that is, male and female.[89] Creatures born of the earth are 'born from the same', that is, asexually, of a single source. Followers of historical dramas are familiar with regal obsessions about lineage and family loyalties (Shakespeare's plays, and historic dramas such as *The Tudors* and *The Borgias*, and fantasies such as in *Game of Thrones*). Lévi-Strauss argues that the myth concerns our inability 'to find a satisfactory transition between this theory and the knowledge that human beings are actually born from the union of man and woman'.[90]

We can draw myths of autochthonous origins into speculations about artificial life, which arguably rework the ancient conflict between being 'born from different or born from the same'.[91] People who make or seek to create artificial life metaphorically fashion life from the ground, or at least they participate in that legacy. Fabricating life is a controversial pastime, and one whose controversies span long before robotics, cloning, genetic manipulation and artificial life. The ancient conflict anticipated the modern putative clash, fusion and reconfiguration involving life and its simulation, the natural and the artificial.

As if awakened from subterranean slumber, autochthonous creatures typically wander and stagger, as do zombies. Creatures emerging from the ground with impaired mobility provide recurrent themes in automata narratives. Many quasi-organic creatures so appear: Tolkien's Orcs, Jules Verne's Morlocks, and Ray Harryhausen's stop-frame skeletons. Even if cyborgs do not all come out of the earth, literally, like the limping Oedipus, they stagger. Think of C3PO's stilted walk in *Star Wars*, the monstrous, transmuting robotic machines in *Transformers*, Gerry Anderson's barely animated marionettes in *Thunderbirds* and the CGI characters that simulate marionette movements in *Thunderbirds are Go*.[92]

On the one hand a so-called 'digital native' or 'digital autochthon' is someone at home in the digital world. Of greater interest, perhaps, is that the quest for digital automata revives a long-standing myth that we are never entirely satisfied with the autonomous emergence of life from life. In keeping with our will to control, we want to make life, to see it emerge from inanimate matter, tantamount to a return to primal origins in the earth.

One major point emerges from this discussion relevant to the Anthropocene and geosemiotics. We humans are of the earth and exercise this affinity by various means, not least in our excursions into the world of the artificial. In so far as we remake and replicate ourselves and our world by synthetic means we are engaging in earth practices, continuing a long tradition of engagement that shows we are always and already of nature. Whatever we do, we are of the earth. It is no wonder that the earth bears the scars of our presence.

Subversives

I wish to follow the mythic thread through to my fifth and final category of Anthropocene response. Imagine dubious practitioners of subterfuge and perjury as among the readers and writers of the Anthropocene. Thornton and Malhi, who I referred to earlier, invoke the desirable characteristic of Anthropocene humanity as that of the *trickster*. I think this is where Anthropocene theory becomes most interesting, and relates to design and creativity. Thornton and Malhi think the trickster has been neglected in the traditional ecology literature. They explain the trickster's character: 'The Trickster is not a devil but an amoral instigator of transgression and transformation of the existing order, pushing it toward something new.'[93] I think the character of the trickster also provides a powerful mythic connection with the ground. For example, the marionette Pinocchio, who entered the ground and emerged again, was a trickster, a dealer in untruths, whose lengthening nose would give him away. There is a further connection between the trickster and ground via the myth of Loki. The god Loki belongs to the Nordic and Germanic mythic pantheon.[94] The main story is that Loki was tied in an underground cave in the company of a torturing serpent. Loki's occasional writhing causes earthquakes.

Trickster is also involved in commerce. Like other trickster characters, such as Hermes and Coyote, the god Loki also plays a central role in the exchange of goods, as explained by modern literature scholar Stefanie von Schnurbein.[95] Loki has several attributes associated with trickiness:

> Loki brings misfortune upon himself and the other Aesir [i.e. gods] with his clumsiness, haplessness, or malevolence, he always redeems himself by dint of his cunning, his magical capacities, or his eloquence.[96]

The psychologist Carl Jung devotes many pages to the elaboration of the trickster. The trickster is the archetype that crosses boundaries and that denies categorization. He is also a potential thief, and a possible liar in accordance with Szerszynski's identification

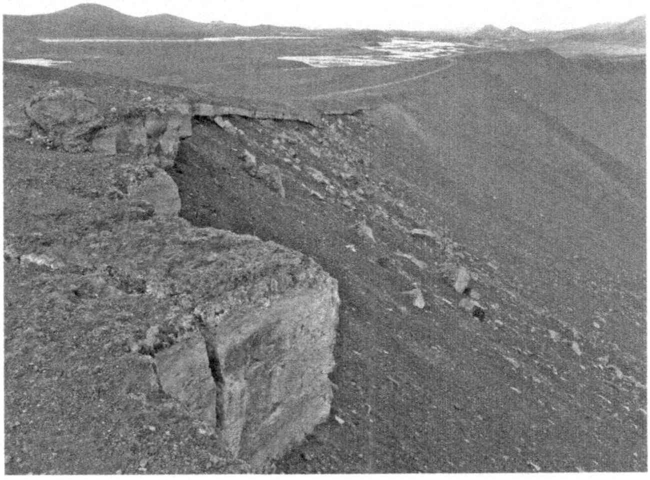

Figure 6.3 Landscape fractured by underground forces, residence of Loki the trickster god. Rangárþing ytra, Suðurland, Iceland.
Source: author.

of the perjurer to which I referred earlier, though as the trickster is a cunning deceiver you can never be sure.

The trickster persists in the digital age.[97] In Chapter 4 I explored biohacking. Some of the techniques of digital activism have moved into the realms of surveillance and espionage by state instrumentalities. Digital activists such as the group LulzSec belonged to a more innocent time. They operated as digital tricksters.[98] According to its manifesto, LulzSec was an 'organization' committed to online hacking because it found it entertaining:

> You find it funny to watch havoc unfold, and we find it funny to cause it. We release personal data so that equally evil people can entertain us with what they do with it.[99]

The trickster–hacker is often portrayed as the maverick, the lone wolf, the wandering coyote, but recent encounters with hacker culture indicate the importance in all human endeavours, at whatever side of the law, of networks of agents, subject to their own rules of solidarity, and dependent on a host of actors: 'There are peons and lulz lizards; trolls and victims.'[100] It seems we are all tricksters, or at least actors in their play.[101]

Similar moves are in play in relation to the environment. *Urban hacktivism* borrows from the digital hack idea and addresses citizen action to prevent unfavourable planning decisions.[102] As a more earthy example, *guerilla gardening*[103] involves citizens planting vegetables on grass verges or spaces neglected by local authorities. The trickster label might be applied to forms of activism on the spectrum from protest, to civil disobedience, to *eco-terrorism*. An article on digital environmentalism by a team led by A. D. Thaler shows how social media blogging can be used to counter misinformation and pseudoscience, inform people about the environment, and bring people together.[104] They make no reference to environmental hacktivism, but do offer a warning against passive *slacktivism*. The latter conveys the sense that you contribute to a cause but do little more than join a social media group, or click the *like* icon on a posting about the environment. They also describe the opportunities for citizen science, such as gathering environmental data and raising awareness of environmental issues.

Returning to the theme of disruption, we see that the trickster personifies the agent who subverts categories.[105] The trickster is in the dubious position between unity and individuation, familial care and harsh legalities. Thornton and Malhi draw on the trickster's 'transforming capacity, unpredictability, clever, resourceful, deceptive, rule-breaking behaviour (sometimes combined with positive hero qualities) and unanticipated cosmic effects (good and bad) from this behaviour'.[106]

As environmental researchers, they favour the Raven who features in North Pacific and Asian mythologies as exemplar of the trickster. The Raven is witness to and source of many earth changes. He pushes boundaries, and steals and redistributes water and fire. He is not a 'god, scientist, humanitarian, engineer or manager, but instead a rogue demiurge'[107] exercising short-term interests and having to bear long-term consequences of his actions. Thornton and Malhi account for Raven in Anthropocene terms: 'In many Native American traditions he begins his existence as a pure white being, only to be permanently blackened by his own misadventures with fire and its sooty, hydrocarbon emissions.'[108] In poetic vein, they relate Raven to 'carbon's metamorphisms, cycles, fluxes and reactions'.[109] They advocate for trickster as the metaphor for Anthropocene humanity:

> Ultimately his transformative, relational, human and other-than-human nature renders Raven a more sympathetic, systemic, ecological and supra-earthly figure than humankind. Although he may lack a moral compass, as a protagonist Raven shows us how to live and how not to live in this multifaceted and unpredictable world.[110]

Elsewhere I have aligned the trickster function with the role of the designer, who otherwise receives scant attention in the texts I have referenced here.[111] I think that most makers, designers, inventors and artists participate in a trickster mindset – as creativity depends on it. Designers are after all sympathetic, systematic, and operate under a necessary fantasy that they can transform the world, and save it.

The 'authors' who read and write the stone book are the moderns, pragmatists, melancholics, children of the earth, and subversives. Alternatively, I could have labelled these five categories: technical optimists, anthropoceneans, grievers, hybrid automata and tricksters. At times, it may seem as though these characters are the perpetrators responsible for earth changes that constitute the Anthropocene. At other times, they are the self-styled heroes and redeemers. If nothing else, this cast of readers and writers shows a wider complicity. No one is innocent. Everything is in play, and there is everything to play for – the subject of the next chapter.

CHAPTER 7
NATURAL SELECTION

Unlikely worlds attract video gamers. What could be more artificial, less natural, less grounded, than sitting in front of a computer screen or gazing down at a smartphone? Can alien and detached experiences reveal something about the lived world, including the natural world? According to cultural theorist John Wills, video games 'remind us of how we create, and have always created, "nature"'.[1] He maintains that such games, 'signpost the virtuality of the real',[2] that is, they reveal 'our seemingly endemic proclivities to make over the natural'.[3] Games have moved on since 2002 when Wills's article was published, but the semiotic contact points between games and nature still hold. According to Wills, nature in games provides 'a synonym for danger',[4] with 'nature cast as savage and predatory'.[5] He refers to the political philosopher Thomas Hobbes's (1588–1679) account of humankind bereft of the trappings of civilized society, where we are at the mercy of our natural impulse, which is for war: 'Every man against every man.'[6] To the political philosopher we are under 'continuall feare, and danger of violent death; And the life of man, solitary, poore, nasty, brutish, and short'.[7] So Lara Croft amid a host of other game heroes has to contend not only with henchmen, mutants, machines and weapons, but also with crocodiles, waterfalls, lightening, cliffs, treacherous currents and other perils of nature. She battles the processes of natural selection to survive as one of the fittest.

In this chapter, I will relate nature, semiotics and place through the theme of play, with focus on the high-profile and controversial influence of video gaming.[8] I will refer to other game-like interventions into urban living, such as free running or parkour. Such practices lead us to think about conflict, or the play element of *agon*, contest, evident in many aspects of the life world, including in nature. My argument leads to Peirce's insights into the transformation of signs via an *abductive* reasoning process – a process of sifting evidence. As well as having something to say about the detective element in gaming, I show that abductive reasoning is inevitably game-like and conflictual.

Play and nature

How do video games present nature? I have already alluded to nature as providing antagonists against which battle is waged. But sometimes in games nature serves as a benevolent guide, 'assuming a folkloric kinship between human and animal creatures',[9] according to Wills. He thinks that in video games 'nature thus translates as good counsel'.[10] This function echoes the many romantics who thought we learn from nature. For the nineteenth-century essayist Ralph Waldo Emerson, 'the ancient precept, "know

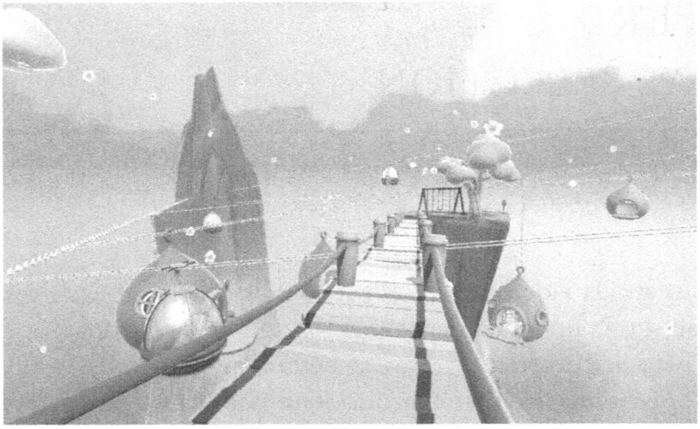

Figure 7.1 Synthetic landscape in computer game by Daoliangzi Zhang. MSc Design and Digital Media (supervised by Jules Rawlinson) used with permission.

thyself," and the modern precept, "study nature," become at last one maxim.'[11] Accurate or plausible representations of natural settings in games can also add authenticity, not just by connecting with nature as a source of authentic engagement, but by presenting recognizable elements and settings for the game play: 'Designers recreate flora and fauna in digital guise to grant legitimacy and coherence to their artificial worlds.'[12] This is common enough in representations in visual media, but unlike a romantic painting or a film, games involve interactions and prescribed causal mechanisms. Wills suggests that games programmers therefore 'resemble Enlightenment philosophers in their treatment of nature as a machine'.[13] Video games also draw on nature as a source of the unknown. Nature's mysteries provide cause for curiosity, but some games also present nature as a source of mysteries to be conquered as gamers follow the 'Colonial impulse'.[14] Nature in games provides a trove of resources: a place to find minerals, hunt game and overcome obstacles to win credits. Some of these challenges are modelled on outdoor pursuits, such as rock climbing, paragliding and survival training. Nature is also a playground. Not only do players frolic and fight within nature spaces, but they also play with putative natural elements, directing the course of evolution in so-called 'god-games' such as Spore.

Then there is the simulation of 'artificial life' in video games, where natural elements grow and evolve, and exhibit animal- and plant-like characteristics, individually or as populations.[15] A kind of techno-nature also develops, with the presentation of animal and machine hybrids and mutants. As well as nature's dangers, its exploitation and decline, some games portray a utopian return to Eden: 'Virtual nature becomes synonymous with romantic sentiment and primordial innocence, its creation a form of digital nostalgia for paradise lost.'[16] Wills notes how in certain genres of video games the 'technological sublime and natural sublime meld together'.[17]

In 2002, many people might have thought of programmers as the main agents in the creation of video games. Now game players know more about the role of designers and artists who bring various talents to the game-design task, not least an understanding of cultural context, the relationships between media, and a ludic and ironic sense to play. The independently produced game also enters the scene. As well as the production of blockbuster games with huge development costs, there are opportunities for freelancers operating with modest means to create, distribute and trial games with relatively modest investments. There is scope for experimentation. Some games exhibit sophisticated approaches to environment, place and nature. Mary Flanagan's book on 'critical play' provides many such examples from independent game developers able to explore politics and play operating outside the mainstream software, board games, theme parks and movie industries.[18]

Furthermore, some games have moved from the desktop and are to be played while on the move, out in the field – in nature as it were. Other games claim to afford benefits like those attributed to wild nature. They are designed to be relaxing, fascinating and therapeutic. One of our students, Hannah Drummond, studied the game *Journey* by TGC as a therapeutic game.[19] The game is set in an atmospheric desert world, is noncombative and arguably therapeutic and affirming. She investigated the gestures and movements of an expert gamer while playing *Call of Duty 4* (an immersive action war game) and *Journey*. She examined signs in the player's demeanour, such as leaning forward, eye movement and teeth grinding to confirm the player's verbal reports that the *Journey* game was in fact more relaxing.[20]

Digital games also relate to nature in other ways. Like fantasy and science-fiction films and other media forms, they keep alive the technological dreams of artificial intelligence, artificial life, autonomous robots and post-human hybridity. Wills says something that hints of this:

> Prolonged exposure to virtual worlds full of digital dinosaurs and artificial life will most likely lead to new ways of seeing, and understanding, nature, especially if so much game play leaves little time for genuine experiences of the great outdoors.[21]

The exaggerated presentations of video games may influence expectations of what nature has to offer: 'Players whiling away their days in the Mario universe may expect from nature a hyper-reality to match the clear colors and textures of Yoshi's Island.'[22] Video games may also

> encourage a tacit acceptance of biomechanical life forms, virtual nature servicing the continual collapsing of boundaries between the artificial and the natural, and the rise of amorphous identities.[23]

I will revisit the prospect that games change our perception of the world further on in this chapter.

Outdoor smartphone games

There are video games that people play outdoors. At the time of writing, the smartphone game Pokémon Go had just been published, to some media excitement, not least due to cases where gamers put themselves in peril chasing after invisible creatures. Though the craze was of limited duration, it is worth looking at the game in some detail. After all, it is an outdoor game, it involves hunting, it nods in the direction of digital augmented reality and it puts the nature–digital relationship to the test in various interesting ways. The game belongs to a genre of multiplayer games and artworks that rely on geolocation. In an article in the *Guardian*, the anthropologist Hannah Gould explained the success of Pokémon Go in terms of 're-enchantment'.[24] The game seems to revive dormant animistic tendencies by populating the world with invisible creatures detected through the magic of your smartphone as you move through the landscape. As explained in many YouTube commentaries there is an economy in play as you capture these creatures (pocket monsters), give them strength, watch them evolve and eventually pit them in contests against other people's monsters. It is a free app, but there are in-game purchases, and advertising, and retailers can attract Pokémon Go players to their physical premises by scattering virtual confetti that lures Pokémons, and hence players, and perhaps customers. It is a consumer-oriented game mysticism inflated to the scale of fully fledged global capitalism – and global-positioning system (GPS) capitalism.

As is well known to location-based game enthusiasts Pokémon Go derives from a game targeted at an older demographic, namely Ingress, by the same company Niantic. It is a precursor to Pokémon Go but has yet to enjoy the same level of success. In Ingress you join one of two globally dispersed 'factions', the Enlightened or the Resistance, and attempt to capture territory by laying down virtual markers, making claims on behalf of your faction, and subverting the attempts of the opposition. The Ingress developers have geotagged monuments and sites around the world that feature in the game play. These sites are then 'portals' that must be joined up to advance the conquest. The Ingress graphics on your smartphone are futuristic and high-tech, which is to say luminous and laser-like. It is a kind of serious war game (a variant of capture the flag). There are no cartoon creatures.

On inspection, Ingress moves close to the spectre of a divided world and real-world conquest. It turns the globe into a game board. This is even more poignant, or sinister, as Niantic was a spinout from the global company Google. Controversy erupted in 2015 when people discovered that holocaust museums and ex-Nazi concentration camp sites were included as portals among the historic monuments.[25] As yet there are no laws preventing anyone from placing a virtual marker anywhere on the surface of the earth, including private property, national parks, areas of environmental sensitivity or restricted areas. A virtual geolocated object is positioned simply via its grid reference coordinates. There is of course nothing there physically. It is just a map reference that activates your GPS-enabled smartphone if you are within range. Thanks to the crude 'augmented reality' graphics, you may see interactive images overlaid on the screen of your device as the app software deploys the device's camera and motion-sensing features. For a popular game such geotags encourage visitors with cameras to congregate at particular sites, inviting physical trespass in some cases and introducing stress to sensitive environments. The video promoting Ingress tells you, the player, that 'this is not a game'. In this and other respects, the semiotic transformation from sacred memorial to game object, and the geographical specificity of the game cuts close to real-world conflict for some communities.

So perhaps that is one of the superior attractions of Ingress's younger sibling Pokémon Go. It is a similar technology, just as sophisticated, and even uses the same geotagged landmarks (as 'PokéStations') but is obviously quirky, juvenile and game-like. It involves hunting for little cartoon creatures instead of parading icons of world domination. Pokémons are styled on domesticated pets after all – harmless, cute, quirky signs of domesticity that pose no real threat to our existence. By another reading, Pokémon Go's cartoony gamification conceals serious intent. As pointed out by the earliest video-game critics, gamers are being further indoctrinated with the imperative to attach numbers to everything (scores), to monetize, to develop fast response skills show readiness to fire weapons and of course indulge the perpetual presence of contest, one-upmanship and conspicuous consumption.[26] By a slightly different reading, in overlaying the world with game-board signs we are rendering conspicuous what we think of as a game, while underneath it all we are still subject to the truth-denying unreality and unaccountability of capitalism.[27] For critics such as Julian Stallabras and Jean Baudrillard, capitalism is the real game, concealed by the veneer of obvious game-like signs and simulations such as Pokémon Go. Ingress and Pokémon Go invite critics to extend their objections to capitalism's hegemonic game scenarios. These games illustrate and reinforce how the whole of geography and nature fall subject to a Cartesian spatial frame.[28] We subdue and limit the experience of space when we abstract space in mathematical and relational terms as data. The danger of spatial abstraction is that we miss the bigger picture.[29] The Pokémon world is populated by game tokens that inevitably fracture the world into discrete objects ready to be appropriated and colonized. Many of those objects are weird animals, to be hunted and captured, thus amplifying our objectification of nature.

Certainly, the inevitable reduction to objects in Pokémon Go renders my neighbourhood in Edinburgh the same as the place I encountered while holidaying in the South of France where I first experienced the game. Though the augmented setting behind the creature is whatever you see through the lens of your smartphone, there is no regional variation in the graphics. I understand that if you are near water then you are more likely to capture a fish (Goldeen), but it is the same fish whether you are by the Mediterranean or the Water of Leith. The streets also look the same as your over-scaled avatar jogs across the two-dimensional Pokémon map.

Reader comments attached to Hannah Gould's *Guardian* article show the widespread reservations many have about such games. The immediate criticism targets safety. In order to capture a Pokémon you have to knock it out by swiping at the screen to activate a virtual billiard ball. In my short foray into the game I was soon invited to stagger into a busy street as I tossed the ball towards the taunts of a Pokémon 'Raticate'. Then there are data and personal-privacy concerns as the app transmits what your phone's camera sees, as well as your geo-coordinates, to the company server. Some critics also express reservations about anything so brashly commercial, popular and of course digital as Pokémon Go. For others, it is a good way to get otherwise inert game addicts moving, exercising and into the world. For others, it is a further distraction from the benefits of

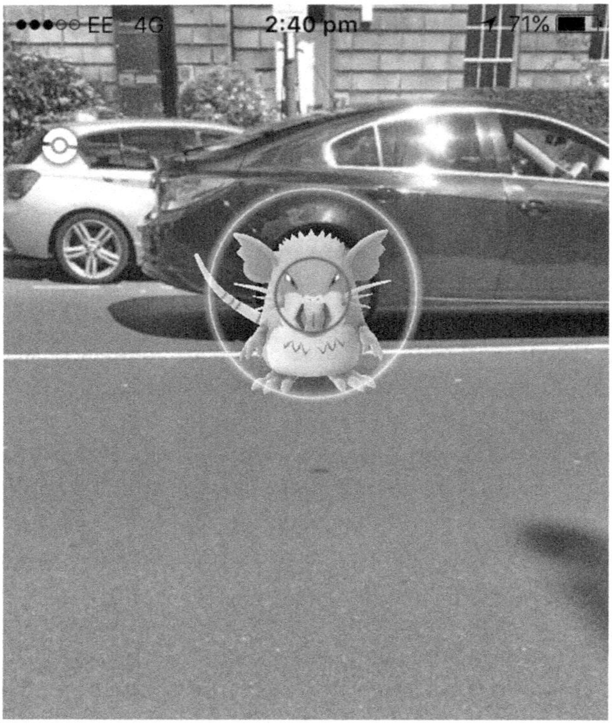

Figure 7.2 Raticate, an augmented reality Pokémon Go character appearing on a busy road. ©2016 Niantic, Inc. ©2016 Pokémon. ©1995–2016 Nintendo / Creatures Inc. / GAME FREAK inc. Pokémon and Pokémon character names are trademarks of Nintendo.
Source: author.

communing with nature. One correspondent 'couldn't help but wonder if our lack of human contact these days, as well as our apparent lack of desire to, you know, WALK anywhere, is leading us to some kind of "Wall-E" future that we're too blind to see'.[30] While taking on board such critique, I think it is interesting to see how such gaming phenomena catch on, and what they imply. I think there are three interesting corollaries to the game's adoption.

First, Hannah Gould's article refers amusingly to someone staring at a real bird on the pavement and wondering aloud what its 'combat power' is. Something similar occurred in my own experience on the weekend that Pokémon Go was released in Europe. While on holiday in the town of Menton in the South of France I aimed my mobile phone to take a picture of a cat in a laneway. Two young men glanced in my direction as they walked by, and I distinctly heard one of them utter the word 'Pokémon'. As was reported in the papers at the time, Pokémon Go provided opportunities for strangers to meet and talk. A day later, the app was available for download to UK subscribers. The people observing me photographing a pet as if it were a Pokémon resonates as a kind of joke. Seeing wildlife and pets as Pokémons, and tourists as gamers, speak to the power of games and the human imagination. Technologies, art and new practices challenge, reveal, transform and even subvert perceptions and understandings of nature and the life world.

It is common for technologies, games and artworks to invite adjustments to our view of the world even when those things are absent. Something similar applies to skateboarding and other putatively subversive urban practices. How does this reframing occur? First, games and other pastimes frame the world of the gamer. Ardent skateboarders see city spaces as ramps and hazards, carpenters see a world full of offcuts to be salvaged for the next job, gardeners see flowers and trees before they notice buildings, geo-gamers think everyone else, like them, is on a scavenger hunt, and Pokémon Go players see cats and sparrows as ready to be captured, scored and inventoried.

Second, self-aware enthusiasts exhibit a countervailing tendency: I can go to places my skateboard will not permit; there are things I cannot put a nail to; it is warmer indoors than in the garden; sometimes I can wander without looking for anything in particular; and look, the bird flies away, the cat purrs and is furry to the touch. The constraints of the digital have the capacity to highlight the properties of the rest of the world, and even enhance our appreciation of what we might choose to identify as the world's power and richness, in particular the richness of what we choose to identify as the natural world.

Third, game play comes to epitomize what it is to subvert the rules. Mary Flanagan develops such insights further in her book *Critical Play*. She states that as games operate as rule systems, they are 'particularly ripe for subversive practices',[31] as people push the boundaries of the game play, and even 'cheat' against the rules. Games have the capacity to subvert, as when we make a game out of a serious situation. They also foreground and bring into relief the idea of political subversion, and operate as the domain of the trickster, who I introduced in Chapter 6.

Nature games

Video games have the potential to alter our view of the world. In this they are in the company of other disruptive practices. Parks, gardens, tree-lined streets, balconies, atria, glasshouses, allotments, bird feeders, green walls, nature reserves, aviaries and zoos are among the most obvious ways that planners, designers and citizens bring nature into the city. But something similar happens via certain marginal urban practices, which by their very nature construct and reconstruct the city as wilderness, bringing the values of the untamed outdoors into the urban asphalt, concrete, brick-and-stone 'jungle'. Such is one of the claims of urban spatial practices such as parkour and its relatives.

Cities play host to a range of spatial activities that are at the edge of civility. Such practices are inconvenient to some, often hazardous, opportunistic, unofficial and occasionally entertaining. Think of graffiti, skateboarding, rooftopping, parkour, free running, begging, busking, sleeping rough, demonstrating and occupying. Such marginal trickster spatial practices appropriate places and city paraphernalia in ways other than their sponsors, designers, legislators and polite civilian users intended. In fact, these practices challenge the concept of intentionality in the design of streets, buildings, parks and malls. Some people, for some of the time and with different motivations, will find ways to subvert the normal, sanctioned and official usage of a thing. They see different affordances in everyday objects. They read the signs of the city differently. They break the established rules of the game. From a semiotic perspective, such subversion adjusts the relationships of the sign otherwise taken for granted.

In an academic article about parkour, Jeffrey Kidder notes how 'individuals appropriate physical space and transform it into something useful from their perspective. Handrails become slides; gridlocked streets become mazes.'[32] The regular significations of the city become something else. He recounts his own foray into the pastime of parkour (PK). After a while, you start to develop 'PK vision': 'Suddenly, the low and wide wooden posts that lined the parking lot jumped out at me. They were about two feet off the ground and five feet apart, and they would be an ideal place for practicing precision jumps.'[33] For the neophyte traceur (parkour practitioner) new relationships between objects, and their affordances, start to assume prominence. I assume such transformations come with acculturation to any urban practice: the jogger who starts to see the world in terms of uninterrupted paths and circular routes, the cyclist who assesses the city in terms of gradients and congestion and the rough sleeper who thinks of the city in terms of shelter, security and invisibility. For Kidder, 'traceurs are remaking the city – turning bland structures like ledges and walls into objects of play. And this play is not only enjoyed by traceurs; it is consumed by others as well.'[34] Such transformations constitute play in several senses. Walls signifying boundaries come to signify obstacles as part of game play. Wittingly or not, traceurs play about with signs. So those of us who know about parkour, or who see its representations in film and online, or chance upon groups of traceurs, reflect on parkour's spatial implications, particularly those designers among us. As such, we are all under the influence of parkour, along with many other marginal urban practices. Such remakings and re-signings of the city are not just foibles in perception by

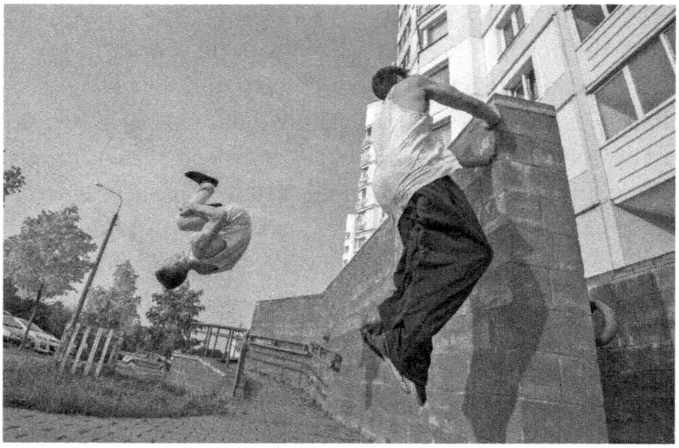

Figure 7.3 Parkour. Urban furniture as objects in nature.
Source: Shutterstock.com/Belarus.

individuals, but are a feature of the shared perception of a group. For Kidder, 'PK vision does not solely reside in the eye of the individual traceur. It is a collective process that comes about as traceurs interact with each other',[35] and with us.

Kidder relates the practice to online videos, which are after all the main source of inspiration for the contemporary traceur, whose 'imagination is inspired from images and texts circulating within the virtual world'.[36] Such marginal urban practices are in wide circulation and colour the way many of us see the environment. As for geospatial gaming, they also bring into question our usual ways of seeing. The city is after all made and remade through many perspectives. No doubt there are other perspectives yet to find expression, and yet to provide overt influence on the design of cities. On the face of it, free running, buildering, rooftopping and the like re-sign walls, roofs, parapets, street furniture, ledges, railings and structures as though mountains, rocks, caverns, streams, boughs, trunks and other elements of nature to be scaled, jumped over, swung from, jumped off, balanced on and vaulted over.

Such activities have their origins in outdoor nature pursuits. Cultural studies theorist Michael Atkinson traces the history of parkour to its foundation in training for jungle combat, and the idea of the obstacle course, as developed by French Vietnam veteran George Hébert (1875–1957) – a deeply respectful pursuit of fitness training in the outdoors.[37] In the absence of balance beams, ladders, rope swings and organized obstacles, suburban fitness enthusiasts resorted to the materials at hand. This was the approach of the founders of the modern parkour movement, Raymond Belle and David Belle. The practice seems to have inherited Hébert's passion for nature. For the Belle family, 'their use of concrete and steel city spaces jibed well with Hébert's philosophy of immersing oneself in one's immediate physical/natural environment to gain a deep phenomenological awareness of it'.[38]

Atkinson identified the persistence of this positive orientation to nature in his study of contemporary traceurs.[39] There is an element of the radical flâneur (as expounded by Walter Benjamin) in the traceur. There is also an ideology in play, of challenging the status quo, the impersonality of the city, its restrictions and exclusionary zonings, inequalities, commodification of life and environment, and pernicious promotion of capitalism. The return to the natural conflates with rebellion against industrial production and consumerism. Atkinson also references Thoreau's *Walden* as a response to the putative stress of city living. The traceur is like Thoreau, the inhabitant born and bred in the city who flees to the woods to experience its full pleasures and pains, where he can 'suck out all the marrow of life'.[40] But here the innovation is not to retreat to the countryside, but to treat the city as if it were nature.

In keeping with this legacy, parkour rejects organization, and offers itself as an uncompetitive sport, or post sport,

> one that subverts modernist ideologies and practices outright and is one in which corporeal dichotomies between the sacred and profane, the raw and the cooked, the civilized/socialized and the primordial body are challenged through athletic movement.[41]

For Atkinson, 'postsports are at once moral, reflexive, community-oriented, green, spiritual, anarchic, and potentially eros-filled physical cultural practices.'[42]

As I have already proposed, the clue to the potency of such marginal urban practices resides in their ability to render strange the familiar environment of the city by transforming its sign systems. According to Atkinson, 'urban traceurs argue that their movements in the city only appear as strange because the "natural" environment of their city is, in itself, strange'.[43] Here Atkinson appeals to Heidegger's concept of *poiesis*. He says, parkour 'brings forth and reveals, through its obvious strangeness there, the fundamentally dominated nature of city life by late modern ideological codes and practices'.[44] Parkour claims a space among other physical and meditative practices by which adherents move into the zone, and enter the flow.

There is indeed something strange about these urban practices. Parkour and other modes of street practice are distinctly urban. They reflect urban problems, and offer responses that enlist the tools and elements of the urban environment. They offer no plan for retreat to the woods, plant gardens, or in other ways remedy the conflict between nature and the city. And yet, these marginal urban practices (parkour, etc.) present as products of nature, born of a love of the woods. Such practices seem to hark back to our primitive place in the wilderness, and try to recapture it by urban means – using the language of the city. Greening the city, building garden bridges over the Thames[45] and other 'bourgeois' initiatives do not have a monopoly over nature in the city. There are other responses, many of which implicate digital technologies as outdoor game elements, as media for the promotion of such pursuits and as means to social organization and the sense that everyone has an opportunity to contribute to the life of the city.[46]

What I have been describing as a play in the transformation of signs also includes the play of metaphor. To see one thing as another is a basic semiotic and metaphoric operation. It implicates imagination, play, exploration and creativity.[47] These points will be addressed further in Chapter 8.

Play and nature

One of the virtues attached to less-energetic geolocative games such as Pokémon Go is that, despite their risks and disadvantages, they encourage game addicts to get outside and exercise. Whether addicted to video games or not, children are frequently told to 'go outside and play'. That is so they do not annoy the cat, smear chocolate on the carpet and so that adults do not fall over them. It is also to get them away from the lure of television and game consoles. But 'go outside and play' trips off the tongue more readily than 'come inside and play', or play in the garage, tool shed or the Land Rover. It is probably safer too, though outside is best thought of as a garden, an enclosed safe area. In fact, the model of the outdoors that many people favour is not some untrammelled wilderness, but the safer, in-between zone between inside and really 'out there' (the wild, elusive, pre-inhabited natural other).

Thanks to gardens, sophisticated adults easily associate outside with play. For landscape historian Jonathan Conlin the long-gone Vauxhall Pleasure Gardens in London incorporated 'elements of masquerade, chinoiserie, and other exotic fantasies that transported visitors to new realms of fancy'.[48] The presentation of such fantasies indicates a transposition of exotic, playful signs into the city:

> Sudden contrasts of light and dark, familiar and strange, pleasure and danger that would have seemed deeply unsettling anywhere else became a source of excitement and wonder.[49]

For people brought up in temperate climates, the natural world at its best presents as bucolic, as a vast parkland of pleasure and play.

Video games are similar to pleasure gardens in some respects. I hope I have shown that play serves as a useful conceptual intermediary between digital technologies and nature. To ask how smartphones and other mobile media assist or hinder our relationship with the natural environment is to ask how they help or hinder play. Music provides similar connections. In writing about music in everyday life, which these days involves smartphones and other mobile media devices, Tia DeNora says, 'to play is to dream in the medium of action'.[50] She adds,

> play furnishes the lifeworld with opportunities for action, with things (roles, riffs, possibilities, personae, scenarios, postures, action chains, styles) that one can play, replay and play over and play around with, together in ways that access forms of experience and ways of being in the world.[51]

Play is about 'making a place'. Technologies inevitably come into play in this place making. DeNora says: 'If play is engagement with the world, then the features of the world can be understood in the broadest sense as toys.'[52] She does not assert this but perhaps the digital accessories we urbanites occasionally carry with us when we go outside and play are as much a part of the 'outdoors' as the trees, flowers, mud and sunshine, and they signify as much through their chameleon-like multipurpose apps.

Urban contest and interpretation

Trending geolocative smartphone game apps are one influence among many in our framing of the environment. The ubiquity of gaming does have an influence on the way we see the world. At least, it brings the play aspect of all life to the fore. Part of that gaming aspect comes down to the idea of interactivity. Interactivity presents as benign and pleasurable, but it is a tamer version of something that has always been in the world of nature and of play, namely *contest*.

I am drawn to the word *agon* (contest) as one of the game patterns identified by Roger Caillois in his book *Man, Play and Games*.[53] Theorists of computer game design have adopted Caillois's categories. There are games of vertigo (rushing about, spinning, jumping), mimicry (dressing up, avatars), chance (rolling dice, taking risks), and agon (battles, fights), and of course any game may have these in combination. *Agon* is

> a question of rivalry which hinges on a single quality (speed, endurance, strength, memory, skill, ingenuity, etc.), exercised, within defined limits and without outside assistance, in such a way that the winner appears to be better than the loser in a certain category of exploits.[54]

I think the case for *agon* in interactive computer gaming is easy to make, as it is for nature. Nature 'red in tooth and claw'[55] is certainly a site of contest, characterized too by Darwinian natural selection and competition between species: 'The survival of the fittest'.[56] The 'urban jungle' presents as a similar site of life and death contest.

Architectural theorist Wendy Pullan draws attention to the role of *agon* in understanding urban environments.[57] From the ancient term *agon* we derive the words *agony* and *antagonism*. By her reading, dynamic, interesting cities, and the architectures of which they are composed, survive and thrive through their participation in the realms of *agon*. If nature is a site of contest, then so are cities. Historian and game theorist Johan Huizinga argued that play is ubiquitous in all human affairs,[58] and the all-pervasive presence of contest supports this. The ubiquity of video games, geolocative games, parkour and other ludic urban pastimes provide further confirmation.

Pullan points out that the agonistic play that underlies such urban pastimes is not subject entirely to rules. Identification of the protagonists, their differences and causes is fluid, contingent and subject to the workings of interpretation, and rightly exercised, debated and worked out in public life. So, an architecture that provides space for public life is crucial for the working of *agon*. She states:

> Place, by being structured in everyday activities rather than regulatory systems, can begin to open a territory where the necessary flexibility of *agon* can exist, with all of its paradoxes and ambiguities.[59]

Agon is a play function, amplified and brought into sharp relief in some of the marginal urban practices I have alluded to so far, including video gaming.

Pullan refers to interpretation and the philosophy of Hans-Georg Gadamer who makes play a central motif in his explanation of interpretation.[60] This was my own starting point for understanding play,[61] and I would like to expand hermeneutics for a few paragraphs for what it reveals about play.

The interpretation of a word, utterance, text, sound, essay, drawing or other collection of signs is a matter of the interpreter's experience, background and the interpretive community in which she or he operates, and conflicts and contest can arise due to these differences in background. But this base of experience is not static. Drawing on Huizinga,[62] Gadamer indicates the inevitable to-and-fro movement by which we interpret a text. We approach an interpretive situation with a certain set of expectations by virtue of our background. If we are open to the text, then it speaks back to us and renews, revises and may even modify our position whether we agree with what the text says or not.

The act of interpretation as a repetitive to-and-fro movement also exemplifies play. In affirming the importance of play, Gadamer is keen to point out that play is not really under the control of the players. In this observation he implicates nature. Apart from the obvious examples of contestants in tennis, team sports, chess and music, he indicates how we invoke play metaphors in talking about nature: 'We find talk of the play of light, the play of the waves', and 'the play of gnats'.[63] He extends such examples to 'the play of gears or parts of machinery, the play of limbs, the play of forces'. He then adds: 'The movement backwards and forwards is obviously so central to the definition of play that it makes no difference who or what performs this movement.'[64] A few paragraphs further on in his section on play, he asserts that there is no difference here whether we are speaking literally or in terms of metaphor. So, play is everywhere, including in the mobile and dynamical aspects of nature. It is worth quoting an extended passage here:

> The fact that the mode of being of play is so close to the mobile form of nature permits us to draw an important methodological conclusion. It is obviously not correct to say that animals *too* play, nor is it correct to say that, metaphorically speaking, water and light play *as well*. Rather, on the contrary, we can say that *man* too plays. His playing too is a natural process. The meaning of his play too, precisely because – and insofar as – he is part of nature, is a pure self-presentation. Thus in this sphere it becomes finally meaningless to distinguish between literal and metaphorical usage.[65]

In this passage, we have a conflation of play as metaphor and play as an actual experience in our life world, a distinction often characterized as the figurative versus the literal. The distinction is moot according to this passage. The fact that both the elements of nature and human beings indulge in play helps dismantle further the distinction between nature and human beings, if not between nature and artifice. There is also a challenge here to the agency of play. Who or what is playing?

According to Gadamer, we get caught up in play most evidently when 'the player loses himself in play'.[66] I think there are many lessons from Gadamer not only about interpretation, meaning, nature and agency but also, as I have elaborated elsewhere, about the character of video game play, in terms of repetition, levels, progression, realism, sociability, ethics and meta-game play.[67] Is there anything else to be said about play beyond hermeneutics? From my point of view the main value of Peirce's semiotics and that of his followers is that it reinforces the place of non-human agency in the processes of interpretation, and hence what we think of as nature.

Game semiotics

I have alluded only in passing to the role of signs in the articulation of contested places. As I have indicated already, semiotics is the study of communication from the point of view of signs, and how we interpret signs to derive meaning. Many of us working in architecture and the arts are exposed to semiotic theories via the work of the Swiss linguist Ferdinand de Saussure.[68] The theories of structuralism and poststructuralism developed from his work. Peirce was a contemporary of Saussure, but Peirce and Saussure were unaware of each other's work. Peirce said little about play directly, though he addressed metaphor and creativity,[69] from which we can make certain inferences about play, and hence nature, computer gaming and contest.

As I attempt to link nature and play, it is worth teasing out the play element in Peirce's thought as well. Nature is there, but where is play? To address this question, it is worth presenting some further basics of Peircean semiotics.[70] Peirce proposes that any sign relation, or situation, consists of the sign, the object to which the sign refers and the so-called 'interpretant'. First, the sign is the mark on a page, the drawing, the pixel image on a computer screen, the sound, the park bench, bollard, ramp or anything that is available to be interpreted. A word on the printed page is an obvious sign, as is a diagrammatic icon on a computer screen, for example, a stylized manila folder, a logo, or an image of a Pokémon character. Second, the object is the thing to which the sign makes direct reference, such as a computer file in the case of the folder image, an instance of a Pokémon creature, or a concept such as boundary, security or well-being. Third, scholars attribute the word 'interpretant' only to Peirce's philosophy. For Peirce, the interpretant is the effect on the person interpreting the sign. In more everyday terminology it is the meaning of the sign, but 'effect' is also a useful term as it implies that meaning involves practical use, as a call to action. Peirce's idea of the interpretant also admits feelings into the range of effects, which enables him to give an account of the way signs are deployed in the arts:

> This 'emotional interpretant', as I call it, may amount to much more than that feeling of recognition; and in some cases, it is the only proper significate effect that the sign produces.[71]

He uses musical performance as an example, referring to music as a sign: 'It conveys, and is intended to convey, the composer's musical ideas; but these usually consist merely in a series of feelings.'[72]

Finding and capturing a game character (e.g. a Pokémon) provides a sense of pleasure, disappointment or tedium, which is part of the sign relation. Peirce says little about where that feeling comes from, implying that it is put there by the author or game developer ready to be picked up by the player. From a hermeneutical perspective, the feeling we get is part of the interpretive process, and derives from the player's prior interpretive experiences, the player's background and expectations. Whatever the performer 'intends', the interpreter listening to the music may experience something different, and in the realms of art such differences are themselves sources of pleasure. In the case of a game, the play is not just the movement of tokens in the game that is the source of our feelings about the game, but the interpretation of the tokens, the game environment, what brought us to the game in the first place and the situation of the playing.

For Peirce, almost anything can serve as a sign: a cloud, a picture of a cloud, a leaf, a footprint, a pictograph, a signature, a car horn, your initials, a word. The thing that the sign refers to can be a physical object, a class of objects, a term, an idea: my favourite apple tree, trees in general, freedom, sustainability or another sign. The interpretant can be whatever the generator of the sign intends, or that is established by some causal connection, or occurs by chance or is invoked via the interpreter's choices as to the meaning. The vast range of possible sign conditions led Peirce to develop his complicated classification system of signs (icons, indices and symbols among others) as I outlined in Chapter 3.

Contrary to Peirce, the structuralism of Saussure bypasses much of this terminological complexity by insisting on the primacy of just the signifier and the signified, a formula that led Jacques Derrida and other poststructuralists to observe that the signifier–signified relationship is indeed complicated, involving indeterminate chains of signification.[73] Any classification system we construct around the idea of the sign is bound to come unstuck at some point, leading to inconsistencies and even paradox. This leads some to think of meaning as elusive, and the derivation of meaning as something of a game-like quest. From a poststructuralist position, the contest, *agon*, or *aporia* (as Derrida characterizes it)[74] arises so often from the priority given to the signifier over the thing signified. By extension, an opposition arises between one term and its opposing term, a duality within which there is always a privileging. I would be going too far astray here if I were to expand further on Derrida's position; suffice it to say that structuralism and poststructuralism provide ample support for the ubiquity of contest and play.[75]

Game detectives

For me the most compelling and game-like aspect of Peirce's insight is how he identifies the indeterminate process by which we derive one sign from another. As I have explored in Chapter 3 he calls the most creative aspect of this process *abduction*. Abduction is where the interpreter exercises creativity. Here I draw on a helpful book by Douglas Anderson, *Creativity and the Philosophy of C.S. Peirce*.[76] From a Peircean perspective, signs are on the way to becoming propositions in language. We translate one sign into another by reasoning. Peirce explains the reasoning process in terms of Aristotelian logic, and uses a bag of beans to do so.[77]

Following the format of the syllogism, the usual *deductive* process starts with a rule, such as 'All the beans in this bag are white'. Then the reasoner, unable to see the contents of the bag or where her hand is located once in the bag, takes some beans and declares factually 'these beans are from this bag'. This is an uncontroversial sign, but as we are using the terminology of the syllogism it is also a 'case' or specific observation. Without even looking at the beans the reasoner can infer by logical deduction, 'therefore these beans are white'. That last declaration is a 'result', an elaboration or translation to a new signification: not just beans but *white* beans. Assuming the rule and the case are correct the result follows with certainty by logical deduction, and there is no need even to look at the beans for confirmation.

A different process, *induction*, involves making inferences from a set of cases. It is about deriving a rule from observations. So, you might take a few beans from the bag and declare, 'these beans are from this bag'. You then notice that the beans are white. If that then leads you to the rule that 'therefore all the beans in this bag are white', you have made an inference by induction. That is risky, as it could just have been by chance that you selected white beans from the bag, which could contain beans of other colours. The confidence that 'all the beans in this bag are white' is in fact a rule enhanced with the collection of further cases, that is, by taking further beans from the bag. This is the usual model of induction proposed as a model in experimental science: you derive rules from lots of cases.[78]

Abduction takes an even greater leap of faith. You may start with the rule that 'All the beans in this bag are white' and then encounter a pile of white beans sitting on a table, 'these beans are white'. If you then infer that 'therefore these beans are from this bag', then that is an abductive inference. Of course, the white beans could be from somewhere else entirely. So, the inference takes an even bigger leap of faith than induction does. In fact, if we deviate from the strictures of this particular syllogism, then it is likely that there are many possible rules in operation. There could be many different bags available with different coloured beans and in different colour combinations, and the results observed may be varied and ambiguous. So, knowing which beans come from which bags will involve some investigation, the gathering of evidence and weighing up the likelihood of different propositions about where the beans come from. This is the nature of detective work, medical diagnosis, scientific discovery and most interpretive tasks requiring skill. It is also the nature of much gaming, including puzzle solving, mystery games, chess and poker. Abduction captures this process nicely.

Abduction is evidential reasoning. Think of games in terms of inferences that require the player to read the signs, formulate hypotheses, collect evidence then discard one possibility in favour of another and backtrack to earlier propositions. The player learns, remembers and forgets from one move to the next. Peirce's semiotics is useful as it includes signs other than those in spoken and written language, and this makes sense in investigative work. After all, a scientist, explorer, diagnostician or detective is not scrutinizing texts so much as substances in test tubes, the flow of the river, the microplastic particulate count in animal digestive tracts, lumps on the skin or blood stains on the carpet as evidence.

The Pokémon player is looking for virtual confetti, groups of fellow players and for signs in the landscape, though it is also like a game of collection and endless repetition. More compelling than Pokémon are mystery adventure games. I know the early game *Myst* and its successors best. The games are set in a naturalistic world (though with a strong fantasy element) and the player must pick up clues to solve discrete puzzles. There are also larger-scale puzzles to solve, a story to be uncovered and rewards to be garnered to progress from one level, island or region to another, with the eventual revelation of a complete story, including its various endings. Characters may appear from time to time to set the scene, but in general the player operates alone (or in a group with a multiplayer version) and traverses the world as if the main protagonists have just left the space. So, the game is designed to give the impression that the player is following clues left behind: a burnt-out book, an abandoned laboratory experiment or a wrecked piece of machinery. Many adventure games follow this abductive pattern.

Peirce does not discuss abduction in this context, but it is clear to me that abduction accounts for many aspects of contest and *agon*. Protagonists will disagree about the evidence and the inferences we make from it. Detective stories are as much about the contest between rival inferences among those trying to draw conclusions from (or obscure) the evidence, as they are about a contest between the detective and the suspect. Throw in uncertainty about the rules of the inference, where people cheat and break the rules, and invent new rules on the fly: this is the formula for so-called 'wicked problems'.[79] Problems in the everyday life world are wicked because they do not conform to neat, well-defined syllogistic logic. They are also 'wicked' as they are in the company of disagreement, contest and *agon*. It is no wonder therefore that Umberto Eco, the Peircean semiotician, turned his hand famously to writing detective stories,[80] which explore the interpretation and misinterpretation of signs, the deliberate obfuscation of signs, and reading and misreading of causal plots and the conspiracies behind them. In similar vein, where walkers converse, explore and exercise their fascination while in nature settings, they speculate about causes and indexical signs: why is the river so shallow here, why are there big boulders on the moors, and how far is it to the next rest stop?

Recalling play

I started this chapter with the proposition that the unlikely worlds exhibited in video games have the potential to reveal something about the lived, natural world. I drew on some game theorists who highlight how designers depict the natural world in video games. Some games are played outdoors, including mobile video games such as Ingress and Pokémon Go. My main point here is that such games reveal something about the world, not least by highlighting differences. Certain games and pastimes belong outdoors. I looked at parkour and other urban pursuits. I focused on these and other ludic means by which nature is brought into the city. I considered contest (*agon*) as a key game function that is also ubiquitous in natural and urban environments. These insights helped me to weave gaming into the fabric of semiotics, regarding Peirce and other scholars on the play element in nature and in interpretation. I think the most compelling insight from Peirce into play comes from his identification of the creative, abductive reasoning processes by which we detect, investigate and hypothesize as we work with evidence, likelihoods and probability.

CHAPTER 8
ZOO-SPACE

Imaginative play and computer games derive much of their character from the presence of animals. According to cultural theorist Randy Malamud, animals are 'painted, written, construed, arranged, stuffed, chained, trained, dissected, imagined', and this happens 'with an iron-fisted sense of entitlement and control on the part of the cultural hegemons, that is, us'.[1] In a critical book on the cultural history of animals, Malamud outlines evidence of the 'digital approximation of taxidermy',[2] such as online pictures and video clips. Online you will find a rabbit (called Oolong by its owners) photographed balancing different objects on its head, including a pancake, and even the skull of another rabbit. In the 'Infinite Cat Project' a succession of images shows a cat watching a picture of a cat watching a picture of a cat, ad infinitum.[3] The digital medium of YouTube amplifies our animal-watching tendencies. The human animal's appetite to observe and record other animals seems boundless. We humans put non-human animals in ridiculous situations. Then we laugh at the animals' indifference to how silly and incompetent we have made them look. There is at least one high-profile case where bathers at a beach passed around a dolphin for selfies, a practice that proved fatal for the dolphin.[4]

Like many others, I am guilty of enjoying the company of animals. I will observe a neighbour's dog straining on its lead, pet the occasional cat on a doorstep to get it to purr, photograph farm ruminants munching in a field, take pictures of wildlife, watch nature documentaries and YouTube clips, and occasionally take animal selfies. Taking my lead from a key text by environmentalist and scholar Paul Shepard (1925–96), in this chapter I will examine how integral non-human animals are to the way we humans think about the world. That is, how they influence human cognition, the way we think and inevitably the way we use signs. One of the key aspects of this zoo-semiotics is the way non-human animals present as *different* to us humans.

Almost human

According to Shepard, we humans are fascinated by what differentiates us from other animals. Non-human animals are the closest entity to humans we humans are likely to encounter. Animals are sufficiently like us that they present as caricatures of human beings. In fact, they define what a caricature is.[5] Animals provide ready-made portraits that cartoonists and illustrators find easier to draw and approximate than credible human caricatures. Animal forms are flexible and adaptable.

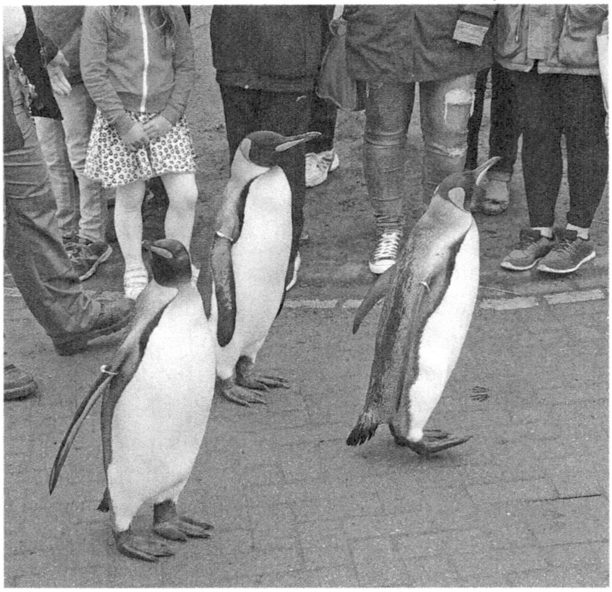

Figure 8.1 Watching emperor penguins at Edinburgh Zoo.
Source: author.

It is common practice to apprehend the otherness of things through the vehicles of humour, absurdity and satire.[6] If nothing else, humour and silliness acknowledge there is something *other* in our encounters. This is one of Shepard's messages in *The Others: How Animals Made Us Human*.[7] From our biological origins as rodents scurrying in the forest to upright homo sapiens, our co-evolution with other animals has forever imprinted in

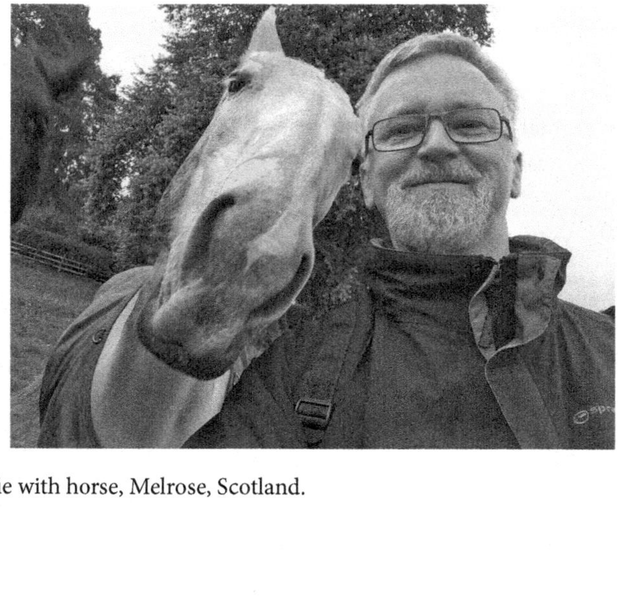

Figure 8.2 Selfie with horse, Melrose, Scotland.
Source: author.

our DNA and cultural memes the important distinction between ourselves and an *other*. Without the sense that there is something a bit like us, but at the same time wholly other, we could not have survived.

Non-human animals are sufficiently different for us to feel we can coerce them into doing our bidding and be whatever we want them to be: entertainer, beast of burden, food source, contestant, comforter, carer, guard or surrogate human. We draw, photograph, shepherd, master, control and consume non-human animals in ways unacceptable if applied to human beings.

Shepard's argument about animals and the *other* hinges on the biological costs and risks for a species with a large brain (i.e. humans), and in turn the concepts of the hunt, as predator and prey, and as participants in a game. Hunting and fishing loom large in this evolutionary tale: 'We are born prepared to assume that opposing forces are "others," and to feel the give the tug of a fish or the musical instrument quicken in our hands as if it were a conscious counterplayer.'[8] That sensibility to otherness gave us the impetus to represent, to communicate, to cooperate, to socialize, plan, scheme and otherwise thrive among the community of all living things. Communication entails the ability to operate at a distance, to read and circulate signs, and bide our time before engaging with the perils of the hunt.

Evidence for the importance of the otherness of animals in human cosmology persists through many cultural forms: animal deities, fables, zoos, mascots, pets, stuffed toys, cartoons, video games, zoomorphic product design (e.g. cars) and attachments to our devices (e.g. smartphones). The design of certain digital consumer devices attempts personableness, that is, a faltering, pet-like compulsion to please. Our fascination with such technologies derives from our engagement with animals as the other.[9]

To identify differences is also to engage with its opposite, similitude, that in turn forms the basis of classification. For Shepard, animals are instrumental in the way we classify and organize information, and think.

Animals and categories

I have already examined the importance in semiotic theories of categorization. Dog, cat, sheep, fox, hen, Pokémon – we do not only classify animals and their surrogates, but animals are also seen as a primary means by which we develop the idea of categories. Shepard argues that animals are 'unusual parts of our environment in that they are radically different from people and from each other'.[10] The fact also that animals move and have differentiated 'anatomical properties' makes them prime candidates for categorization,[11] and categorization is an essential component of thought. So, parents, teachers and the authors of children's books instinctively teach infants about animals. In the process, they teach children about how categorization works: through differentiation, features, properties and affinities, which in turn teaches them to discriminate, evaluate and reason.

For Shepard, it is not just convenience or some immediate usefulness that draws us to animals as vehicles for this learning about classification, but our primordial affinity, and our co-evolution in a universe of animals, traceable to the imperative to hunt, evade, cohabit and survive. Further evidence of this imperative to put things in categories resides in the impulse to collect and sort. Think of collections of seashells, pebbles, eggs, tropical fish, and of course dead beetles, butterflies, stags' heads and hides. Collectors array these formally or informally in musty boxes, on mantelpieces, in albums, museum-display cases and digital archives. Human beings also collect countless pictures of animals and their cartoon representations, with Pokémons, Spore creatures,[12] and other digital collectibles among their number.

Of even greater interest, animals feature as the quintessential breakers of categories, in part contributing to their fascination to us – fascination, again, traceable to the need to sort and discriminate to ensure our ascent among surviving populations and species. Shepard describes the fox (a common trickster surrogate) as such a category denier. Whatever its position in a scientific classification system, the fox presents as a boundary crosser, in that it is a dog (*canidae*) but at the same time has feline qualities (*felidae*); it barks and makes a meowing sound. It roams across the countryside and urban gardens and hunts by stealth. He also notes that 'it is hunted yet not eaten'.[13] Dogs are also in any case occupiers of the boundary. Undomesticated they live on the outskirts of settlements. They live off carcasses, and engage in other unsavoury habits that pet owners try to ignore.[14]

There is a spatial aspect to such animal characteristics and behaviours. For Shepard, 'forms which are themselves at the edges of groups become the focus of accentuated attention and deliberation'.[15] He adds: 'Just as edgeless entities threaten visual chaos, types without borders, ambiguous in their relationships, subvert cognition.'[16] He then highlights our aversion to, or disturbance by, animals that defy classification. They become

> signs of corruption or disarray, or they may be seen as sacred mediators, but in either case the dubious forms create excitement, thoughtful deliberation, and a rich mine of metaphorical ore.[17]

He thinks of humans also as 'edge animals'. We favour the boundary position between categories:

> As apes, dog-heads, yetis, fallen angels, diverse races, and emergent androids lurk in the margins of our identity, our species is beset with a problem of the categorical imagination.[18]

So, animals loom large in the human capacity to categorize, to move in, out and across categories, which is to think, to problematize and to think about the problems of thinking.

Domination and distance

To be in, out or on the border of a category carries implications in terms of power relationships. The categories *human* and *animal* are unevenly matched.[19] Moving to the popular realm of animated feature films, the Disney film *Zootropolis* (2016) reminds us of the impossibility of denying our animal natures. In the film, anthropomorphized animals of different species live in apparent harmony in a city of the future. The animals procreate as usual, but carnivores subjugate their animalistic tendency and the necessity to hunt and consume other creatures. How else could species coexist in such a utopia if they hunt one another? Audiences will accept readily such impossible scenarios in fiction and entertainment. The film is about animals, but, as with any animal fiction, it also concerns being human, and in this case the animal nature within the human. By most accounts the film deals with stereotypes and prejudice.[20] Behind the fascination with animals resides a kind of anxiety, the desire to show that we are different from animals, and to suppress our animal natures.

That we feel superior to and aloof from our animal natures comes through accounts of stories about human origins. According to Vitruvius the classical architectural

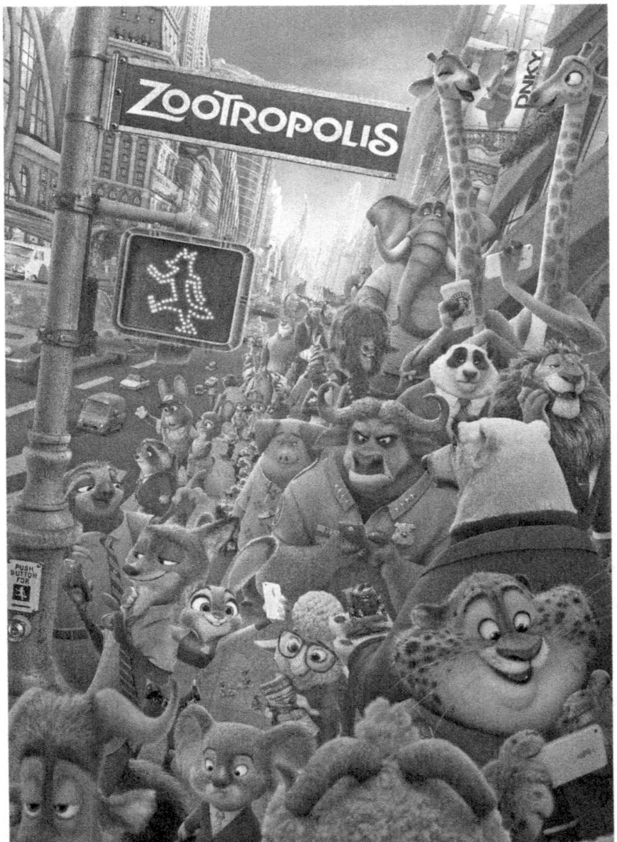

Figure 8.3 Anthropomorphic animals in the Disney animated feature film *Zootropolis* (2016) directed by Byron Howard and Rich Moore ©Disney. used with permission.

theorist, man is superior to the animals in that he can stand upright and look to the heavens.[21] For Shepard, we oscillate between wanting to see our animal inferiors struggle as entertainment (bear baiting, cock fighting, lion taming) and as vulnerable creatures that we nurture and pet. Some representations of animals have big eyes and other human features and behaviours that make them 'cute', so we can incline to parent them as if they are infants. Human beings are also creatures who compete, and we like stories about contest. Cats chase mice, sharks eat smaller fishes, lions chase zebras, coyotes go after *Geococcyx californianus*[22] and hunters pursue ducks and rabbits. As a hunting species, we thrive on stories of the hunt, and our superior position in the hunt. Presumably when human life depended on it such stories transmitted and enhanced our hunting prowess.

So, animals feature in the narratives we construct about human behaviour, relationships and power structures. The Enlightenment reinforced a humanist outlook in which the human was the arbiter in all things, with no god or church above man. Man was at the pinnacle of a hierarchy variously constructed: evolutionary, taxonomic, authoritative, and moral. Man is at the top and animals are below. Gods and angels are nowhere above us, but are at our behest, or our invention, if we want them. Some think that we are now in a post-human condition.[23] I think there are many problems with the term, but post-humanism is a pragmatic orientation. It starts without the presumption of a hierarchy but assumes a heterogeneous and undifferentiated world of experience. We divide up this field for some purpose or other and in particular contexts, much of which could be described as political.[24] Thinking human beings recoil in horror at news reports of people-trafficking, slavery, ritualized execution, and other kinds of abuse. We say that such *inhuman* abuse treats human beings worse than we treat animals, or the way a predatory animal treats its prey. The distinction between the human and the animal becomes political, not only in the field of animal rights, but in the discourse about how human beings treat one another.

Animal decline

Paul Shepard provides an interesting narrative account of the decline in our relationship with and respect for non-human animals. There are five stages in his account. For Shepard, prior to urban settlement, domestication and commerce, nature harboured enchantments lost to us now. Animals roamed as sacred, conscious and individually unique beings. Some even attached themselves to spiritual powers. This is the first stage in Shepard's (deep) ecological account of how we came to subjugate animals, and to treat the gentler ones as pets. He was writing before the internet took hold, but there is ample evidence that pets matter in the electronic age. Not least, live pets have microchips, people talk of robotic pets, people fawn over pets on YouTube, some collect Pokémons and there is the dystopian fantasy that humans might one day become pets for robots according to a reported comment by Apple cofounder Steve Wozniak.[25] Critics of the social media generation also indicate that some people prefer the company of artificial pets to fellow human beings.[26] Keeping animals or their surrogates as pets runs deep in the human psyche.

In Shepard's second stage in our ascent to superiority, and declining relationship with animals, humans took some animals in as members of the household. We 'manipulated their reproduction, and altered their biological natures to conform to human dominance'.[27] So that is about animals for food, labour, guard dogs, and to keep out other animals. These genetically adjusted and manipulated animals diminished the number and diversity of their wilder counterparts. Supported by their human masters, domesticated animals outcompete their wild cousins.

Shepard's third stage is about human infants. The boundaries and securities of the domestic sphere impede the infant's initial craving for contact with the abundance of other life (the natural world, including animals). In Freudian mode, Shepard refers to the trauma of the child's separation from its mother. Cuddly toys are transitional objects that ease the child through this inevitable separation. On the other hand: 'The children who do not seem to require the security of such objects are those who are surrounded by abundant other forms of life.'[28] Shepard theorizes that proximity to animals makes separation from one's mother less frightening, in part as the similarities and differences encountered in the world of animals prepares the infant for such raw events. But now, cuddly stuffed animals and toys compensate the child in the nuclear family for this estrangement from animal encounters.

The fourth stage involves a simpler theoretical transition. Children transfer their affection for cuddly toys onto dogs and cats: 'As the toys had been pets, the pets became toys.'[29] The transition then extends to 'the wild', where any kind of animal gets domesticated in fiction, cartoons and CGI animations: Peter Rabbit, Pooh Bear, Micky Mouse, Simba, Judy Hopps, Nick Wilde, etc.

In the fifth and final stage, we humans extend 'the equivalence of the living domestic pet and the stuffed wild toy to living nature'.[30] Zoos epitomize this transition, according to Shepard. Drawing on an article by John Berger,[31] he indicates that zoos inevitably disappoint us. The animals stare past, and do not do what we expect of them, that is, acknowledge our presence, worth, importance and desire for companionship and solidarity with the wild. Zoos are therefore lonely, melancholic places and not only for the animals. Something similar occurs in the case of close-up documentaries about the lives of animals: 'These are as remote from our lives as the "friendship" of the kangaroo, donkey, tiger, and Pooh Bear.'[32] One of the many interesting aspects of Shepard's account is that it places animal surrogates (i.e. stuffed toys, cartoon animals, robotic pets) at centre stage. They are instrumental in our construction and understanding of wild nature, and help account for any deficiencies in our attitude towards it.

There is a deconstructive play in progress here. Whereas we think of toys as substitutes for animals, Shepard argues the reverse. We think of animals as we think of stuffed toys. Adults, adolescents and infants play with pets as they would with teddies, fluffy unicorns and rag dolls. We play with the elements of nature as we would their synthetic surrogates. In the human imagination, we fashion nature after our synthetic playthings, including our devices, systems, robots and video game characters. Nature imitates artifice, and in this we put into action the translation of signs, which is also the play of metaphor, as I will show.

Uncanny and monstrous

As an important aspect of their contributions to cognition, animals help us define and identify various traits and aesthetic categories. First there is a category of experience readily described as a sense of the uncanny. Non-human animals can provide a potent demonstration of intelligence, or what it is to be intelligent. Some animals even appear wise, indicated by their behaviour, interactions and ability to remember and learn. But most important is their expression. Shepard reminds us that cats never smile. This is what makes Lewis Carroll's Cheshire Cat uncanny. Animals are expressionless, contributing further to their capacity to present to us humans as other – both familiar and alien according to Shepard:

> Deeply committed to the play of facial features and the power of expression, we find the immobile faces of other animals to be suspiciously concealing, or to be the guileless mind of pure, untroubled divinity – transcendent, serene, detached, innocent, knowing.[33]

As they are even less expressive, fish, birds and reptiles present as the most uncanny of creatures.[34] He also says of the fox, 'We sense a being essentially like us and yet beyond us, in the guise of a special wisdom that denies the ambiguity of our own fluid look.'[35] The aspect of each species seems to be 'fixed', 'as if contemplating its own monstrous or wonderful secret, an idea made perfect, as if for our attention'.[36]

He discusses the grotesque display of mounted animal heads in places that celebrate the hunt. Whether mounted, stuffed, kept in zoos or bred as pets, we wait for some recognition as co-inhabitants of planet earth: 'We await a reunion with absent beings on a crowded yet increasingly lonely planet.'[37] To this melancholic quest for recognition – and even forgiveness – from our captive animal siblings add the vast trail of YouTube animal clips. If only the cats, puppies, emus and geese would laugh along with us, but they never do. In this resides the wisdom of the flock, herd, skein and glaring. As if they know something that escapes human wisdom, the strange familiarity of the animal presents to us as uncanny.

If they epitomize the strange, animals also define monstrosity. According to anthropologist Mary Douglas we tend to be disturbed or disgusted by things that are in the wrong place: shoes on the dinner table, saliva on the salad, a snail on the pillow, a mouse in the biscuit tin.[38] Nothing is disgusting in its own right; just the unusual relationships set up by its misplacement. There is little that is disturbing about noses, eyes and mouths, but eyes that look like human eyes on an animal face create something close to monstrous. Animal ears on a human body have a similar unsettling effect. Animals feature in the menagerie of monsters, imaginary and real, that populate our stories. Mixing up animals with different attributes in the same stories also lures us into the realms of the monstrous – Mickey Mouse's friend Goofy and pet Pluto (both dogs we think), and Spongebob (a sea sponge that looks like a kitchen sponge) and his pet snail with the ordinary human name of Gary.

The idea of the monster provides a useful inroad to understanding metaphor that Peirce regards, without elaboration, as a key component in the functioning of signs. If monsters are made up of human and animal parts out of place, then metaphors are also objects in the wrong place (i.e. misclassifications).[39] Mice that have pet dogs, worker ants that make love, fish that talk and caterpillars that read are category errors, which is to say metaphors. Metaphors are part of our language and thinking.

Cute culture

As I explained in Chapters 6 and 7, the trickster is a recurrent type that storytellers associate with elements in nature, including animals: coyote, fox, raven, monkey. These are 'mischievous' animals that try to confound rationality. They make us laugh, and they are cunning deceivers. Animal tricksters populate children's stories and cartoons: the mouse that outsmarts the cat, the wise-guy rabbit, the deranged duck – the hunted becomes the hunter, roles ambiguated and reversed.

Cute is also a trickster category. I have mentioned Shepherd's account of the human tendency to treat animals as if stuffed toys. Bulbous, padded, soft to the touch, yielding: Who could miss the palpable cuteness of puppies and kittens on YouTube, and their mass-produced surrogates in cartoons, video games, on logos, as branded accessories and as soft toys? According to the *OED*, the word 'cute' is an abbreviation of 'acute', meaning clever, keen-witted, sharp and shrewd. In her extensive exploration of *cute* as a viable contemporary aesthetic category, the literature scholar Sianne Ngai notes that sometimes unstressed syllables get thrown away, so 'alone' becomes 'lone', 'until' becomes 'til' and 'acute' becomes 'cute'.[40] In the case of the word 'acute' the removal of the first vowel changes it from sharpness to its opposite, softness. After all, in everyday vernacular speech 'cute' implies something infantile, vulnerable, fuzzy and blobby such as a stuffed toy, a kitten or a baby. Ngai notes that diminutives are born of such abbreviations, as when we affectionately shorten people's names, and then add an extra syllable. As Ngai indicates in the case of 'cute' we have 'cutie', 'cutesy', 'cutie-pie' and even 'cutesy-poo' also in the *OED*. There are several meanings to cute, such as attractive and quirky, but the dictionary also says that cute means 'cunning', which fits the sharpness aspect of its origins in 'acute'. So, the ambiguous, cunning characters of fox, coyote, Puck, Pan, Harlequin, Hermes and other instances of the trickster archetype are also *cute*.

The full title of Ngai's fascinating study is *Our Aesthetic Categories: Zany, Cute, Interesting*. She does not make the trickster connection in that book, but it aligns with what she regards as the ambiguity and duplicity of the cute character with the madcap, hapless, accident-prone zany clown, which she also discusses. She is keen to point out that *cute* is a newish, late-industrial aesthetic category, albeit with a legacy going back at least to Edmund Burke on the sublime: 'Beauty in distress is the most affecting beauty.'[41] We are attracted to people and things that appear vulnerable. She argues that more recently, Japanese kawaii culture (Hello Kitty, etc.) emerged from Japan's loss of power in the wake of the Second World War. Stylized blobby cutes are vulnerable and worthy of everyone's care, like the then-increasingly impotent ageing Japanese emperor. Soft toys can also be beaten up and bashed about, yielding against the inevitable roughness of children's play. Earlier toys were more hard-edged, rigid and ornamental than soft and robust.

Ngai also points to the consumerist aspect of cute things. Cuddly commodities want us to take care of them as they cry out to us for purchase: 'The cute commodity, for all its pathos and powerlessness, is thus capable of making surprisingly powerful demands.'[42] Emojis are also cute, even those with angry frowns, as are Pokémonsters, screen icons and animated GIFs. The consumer world must hint at least of the adorable, the harmless and the cute. It is the way we consumers want nature to be. I take it that cute things are like whimpering puppies in a pet shop competing for our attention. Ngai does not address architecture or space in any direct way, but as I read her work I cannot help but think of the tendency towards the organic and the blobby in biomorphic architecture, though because of its scale it is even more ambiguous than a soft toy. In architecture, the cute transitions inevitably and mercilessly into the monstrous.

Talking with animals

In Chapter 3, the Book on Nature, I outlined the insight of Peirce and others that all of nature is involved in communication through signs. I need hardly reiterate that animals communicate within and outside their own species boundaries, and we communicate with animals. Watching pet owners coach their pets to deploy human speech provides a cute diversion on YouTube. You can train a dog to say 'hello' as a vocalized yawn, or to growl out something like 'sausages'. Animals (non-human) respond to what we say some of the time, but do not talk back in the same way, that is, using our language. A dog cannot tell its owner in words why it wants to be let out to the backyard and what it enjoys most about rolling about in the grass.

In this context, I am drawn to Derrida's essay, 'The animal that therefore I am'.[43] Building on Saussure's theories of signs, Jacques Derrida wrote about speech, talking, writing, meaning and language. He identified animals and their attributes throughout his writings: ants, silkworms, asses, animal sacrifice, suffering, shame, nakedness, etc. In this context, I think of Derrida as Peirce on speed, ludic, subversive and tricky. Derrida joins the chorus of scholars such as Donna Haraway who seek to unsettle categorical certainties, and thereby bring about social and political change.[44] For example, Haraway asserts the cyborg 'as our ontology', a prototypic human–machine hybrid, a monster, a marginal entity that functions as a surrogate for minorities, the oppressed, women, and those without a key stake in the power structures. Derrida's tactic in his essay 'The animal that therefore I am' is a bit different, and puts language at the centre of the critical discourse. According to my reading of Derrida, it is not just that we say animals cannot speak, but it is the way we speak about animals that is open to question and ready for renewal. At least Derrida helps explain our fascination with animals that talk, and machines that do something similar.[45]

I have attempted to show in this chapter how animals feature in cognition, the definition of difference, categorization, metaphor, power relations, aesthetic categories and language. We need only recall via Shepherd how the presence of domesticated animals permeates and defines the domestic sphere. So, our attitude to non-human animals implicates the idea of home, hearth and architecture in general. Our cognitive coupling with animals can be explained in terms of the human's co-evolution with other creatures, our love of competition and the hunt, the animal's availability for semiotic bricolage,[46] the trickster archetype, monstrosity, the role of metaphor, human contest and issues around signs and language. Such accounts already presume our separateness from animals, but they serve to introduce the problematic of the animal and the human. Experiments with robotic pets, and the way we anthropomorphize animals, computers, engineering constructions and buildings also reflect our dealing with the animal in human nature. In this chapter I tried to reinforce the proposition that natural environments are complicit in networks of signification and cognition, and animals are part of that communicative and cognitive scaffolding, as we are part of theirs.

CHAPTER 9
REFUGE

A natural orientation seeks balance, and the restoration of balance. In Chapter 1 I referred to the sense of balance and wholeness that communion with nature brings. I alluded to the potency of the tuning metaphor: tuning in, getting in touch, recalibrating and resetting. Tuning is a musical metaphor. Musicians tune their instruments to musical scales and to each other's instruments. In this context tuning suggests a synchronized vibration, as if we move with the rhythms of nature.[1] 'Attunement' is also a term in phenomenology to account for the human condition whereby we connect with a mood. German scholars use the same word for both tuning and mood. It is *Stimmung* as used by Martin Heidegger. Though Peirce deployed other terms to describe the process, Nathan Houser, editor of one of Peirce's collected works, states that Peirce believed 'attunement to nature was the key to the advancement of knowledge – as it was for life itself.'[2] Houser adds, 'and he thought the power to guess nature's ways was one of the great wonders of the cosmos.'[3]

Architectural theorist Alberto Perez-Gomez concurs with the need to get in tune. In keeping with much of the contemporary literature on nature he aligns attunement with harmony, balance and 'psychosomatic health'.[4] He argues that attunement implicates mood and wholeness. The quest for attunement in architecture is

> a search for lost integrity, wholeness, and holiness; transforming from proportion in the classical and medieval contexts to atmosphere or mood, it becomes a central concept for artistic works in search of meaning, potentially including architecture, of course.[5]

According to the classical tradition, to be in tune is also to participate in a life that is wholesome and good. In summarizing what scholars, classical and modern, assert about *the good life*, philosopher Charles Guignon agrees:

> Representatives of the classical model of the human-nature relationship tend to see the happiest, most flourishing and fulfilled life as one that is "healthy" because it is in tune with nature and at peace within itself.[6]

The term 'nature' certainly attracts concepts of the good. The tradition to which Guignon refers is that of the Epicureans, the Greek philosophers of the garden who advocated a life informed by pleasure.[7] Such discourses resonate with the idea of restoration. The place you go to restore is a refuge. People seek refuge in many places, gardens among them.

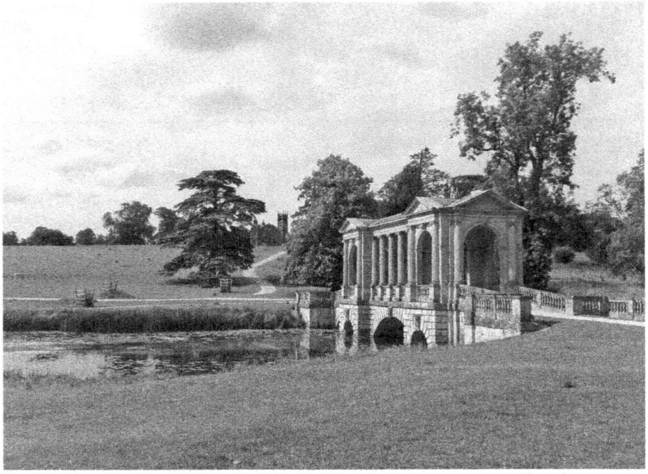

Figure 9.1 Bucolic landscape. Stowe Garden, Buckinghamshire. Gardens designed by James Gibbs, William Kent and Lancelot 'Capability' Brown.
Source: author.

Players also seek refuge in play. Any task that absorbs a person's attention has the potential to offer refuge. Standard meditative practice encourages participants to focus attention, to block out other signs, thoughts, concerns and sources of stress. As I examine in this chapter, key environmental psychologists say something similar about engagement with natural environments. But in this chapter I also present the case that we find many of the positive attributes that we ascribe to nature in other parts of our environments pervaded by artifice, not least in music and digital media. In these and other respects nature imitates art.

Following my observations in Chapter 7 about the unsettled and agonistic nature of play, we may assume that we human beings move into and out of such conditions of restful equilibrium. Perez-Gomez identifies the ancient 'concordia discors, or discordant harmony',[8] that is, the conflict between the elements of nature. I think of this as a venturing forth into a state of unbalance, and a return to a condition of momentary stasis. Organisms balance the risks of hunting and foraging with the risk of staying in their safe place. Much of the nature narrative is bound up in such transitions. A refuge is not a permanent place of habitation, but a temporary retreat.

Theories of semiotics impinge on the nature of refuge and well-being. I have drawn on semiotic theorists who believe that all nature is in communication somehow, from the level of DNA to the full animal, including animal populations and ecosystems. We participate in nature's signs as we are of the same stock, born of the same earth, though we can fall out of that condition. According to this view, health involves renewing our biosemiotic and geosemiotic literacy, getting back to a relationship with nature. There is also therapeutic value in knowing about the semiotic connection of all things, and having 'the power to guess nature's ways',[9] according to Peirce's commentator Nathan

Houser. As I stated in Chapter 1, the ancient philosophy of the stoics built on a belief in such a complex web of interconnections, with the corollary that if we only knew how connected we are to one another and the world around us then we would be less troubled by individual misfortune.[10] As we saw in Chapter 1, semiotics also traces its origins to diagnosis – understanding the health or otherwise of an organism or system by the signs it bears. In some cases, 'semiotic healing'[11] comes to our aid as an explanation of how well-being prospers even when the medical palliative is absent, as in the case of the placebo effect.

In what follows I will elaborate on the attributes of nature as refuge, considering attunement, healthy environments, solitude, media, culture, causality and loss. I argue that once we accept the semiotic character of the natural life world and our place in it, then we must also admit digital technologies, which are after all contributors to the wider semiotic sphere that builds on a kind of 'relational sensibility'.[12]

Attunement

We do not have to attend constantly to whatever we are attuned to. Being tuned in can operate as a background condition, like a mood, of which we are barely aware, though it affects us. Once the radio is set to the station then the listener can forget about tuning, unless the signal drifts, in which case the listener makes small adjustments and re-tunes. Musicians start a performance by tuning their instruments but then forget about the process until the instruments go out of tune. In the case of stringed instruments musicians re-tune between movements or adjust their finger positions or adjust their bowing while playing. To tune in requires a concerted effort for a while at least, and may involve some concentration.

To tune in requires attention: to the instrument, the setting, the mood, the place, the technology, the ensemble and other people. To tune in might seem like something that happens to you, passively blending in, as if just letting things go, but to tune in requires work. Tuning in can be like falling, as when people let themselves be overtaken by the mood of the moment, but even that requires letting go, surrendering to the mood, and that is work.

What is it to get in tune with nature? To tune in to nature is to undertake some work, even if only momentary. 'I do my utmost to attain emptiness; I hold firmly to stillness,' said a poet in the *Tao Te Ching*.[13] According to philosopher Charles Guignon, Taoism 'advocated a "letting-be" approach to worldly affairs, a way of life characterized by attunement to nature and the cultivation of quietude and inner peace'.[14] Attunement to nature requires effort, doing your utmost and holding fast.

Another term for 'to hold fast' is 'to pay attention', to *attend*. As we saw in Chapter 1, to attend means to be present, there, focused. As anyone knows who has tried to keep alert in a difficult lecture or while reading a tedious book, the mind wanders, and you must draw your attention back to the subject, the task at hand, or the topic, and that requires cognitive effort. Attending and re-attending is a bit like adjusting a musical instrument if it goes out of tune, or adjusting the dial if the radio drifts off the station. Practices of meditation typically require participants to focus their attention on something, such as their breathing, and bring their attention back to that if their attention wanders.

Attention and distraction are major players in the busy high-tech world many of us inhabit now, especially when critics want to establish a contrast between live face-to-face social engagement and the world of digital technologies. After observing people engrossed in their mobile devices in a café, one influential commentator complained: 'A "place" used to comprise a physical space and the people within it. What is a place if those who are physically present have their attention on the absent?'[15] The charge here is that digital communications direct the attention of their users elsewhere than the current place and to times other than the present. Smartphones, social media channels, phone calls, video games and email take digital users out of the moment. The complaint is echoed by Perez-Gomez, for whom atmospheric architectural spaces are 'always intertwined with temporality; they are never "outside" time'.[16] So he thinks that truly atmospheric places 'challenge the present-day ubiquity of telecommunications and its supposedly public spaces'.[17] He adds that this is crucial in a world 'increasingly consumed by its obsessions for iPhones and computer screens'.[18]

It is common now to counsel people to leave their smartphones at home when they go for a stroll in the park or the countryside. There is the safety angle as inattentive pedestrians risk collision with one another, motor vehicles, street furniture and other obstacles, and some cities have installed pedestrian lights in the pavement at busy intersections in the line of sight of phone-'obsessed' pedestrians.[19] Whatever benefits nature affords, many people think that phones distract us from attunement with nature. People, places, and digital media channels compete for our attention.

A 2012 press release from a conference run by the British Psychological Society reported the advice that workers should turn off their smartphones to avoid stress.[20] Many online news outlets repeated the report that rapidly became the main source of authority on stress and smartphones.[21] Whatever the evidence for the link between smartphones and stress, some people spend concerted periods away from digital technology to 'detox'. According to an Ofcom annual report on the communications market in the UK, 'digital detox' refers to 'a period of time when a person makes a conscious decision not to go online or use connected devices'.[22] It is also 'an opportunity to focus on offline activities such as exercising, socialising with friends and family, doing housework or homework, or simply relaxing'.[23] Digital detox is one of the headline investigations in this review of UK consumer behaviour in relation to television, telecommunications, post and of course online digital content. About one-third of internet users surveyed had tried to take a break from online activity – a 'detox' period of a few hours to several days. The report says that the positives of the experience 'far outweighed the negatives'.[24] The pluses included feeling more productive, liberated, enjoying life more and feeling less distracted. Smaller numbers reported negative experiences such as feeling lost and anxious.[25]

Part of the argument for release from digital stress revolves around the issue of attention that I broached in Chapter 1. If a person is conscious then they are attending to something or the other. It seems that if you attend for a long period to your computer and some important work task, then you eventually succumb to a kind of 'attention fatigue'.[26] It is likely that the work task also involves pressures of time, and is accompanied by certain challenges that in time deplete your cognitive resources – or it feels that way. From a biological perspective, such work weariness carries certain advantages. It alerts us to the need to change our activity. Continuous and uninterrupted fascination with the task at hand leaves us prone to external threats. So we need to recover from that state. Restoration comes from a change in activity, that is, attending to something else. Then you can return refreshed to the more challenging task at hand.

To what tasks should we turn for recovery? You can gain respite by various means. Environmental psychologists Stephen and Rachel Kaplan refer to the potential of playing games, watching television and other diversions to deliver refuge from work-based attention.[27] But the 'natural world' provides opportunities for directing attention that are less demanding or stressful: such as attending to the texture of the bark on a tree, the sound of the birds, movement of the clouds or a distant horizon. Their strong point is that natural environments attract people's attention. The Kaplans alert us to the 'soft fascination' afforded by plants, earth, rocks, mountains, clouds and the whole sensual panoply of the natural world. As I have already suggested, meditative practices typically advocate such attention: attending, focusing attention, but at the same time doing so in a relaxed state and without urgent demands, performance pressures and other stress inducers. Soft fascination is what turns natural environments into sites of well-being.

Healthy places

Studies connecting health and environment are substantial.[28] The claims that natural environments have a restorative function inform land-use policy. The Scottish Forestry Commission has produced a series of reports outlining why it is good for people to get out into forests and green spaces:

> There is a strengthening body of evidence to support the view that greenspace and woodlands provide the ideal setting to promote health and physical activity.[29]

We can also bring such health benefits into the city. In their research into sustainable cities, Timothy Beatley and Peter Newman present the restorative benefits of nature in terms of people's emotional attachment to places. They maintain that

> Urban environments that are greener, more nature-full, will attract greater interest by residents and help to strengthen emotional bonds to place and community, in turn increasing urban resilience.[30]

They refer to the health benefits of natural environments within or close to cities. People will get out more, walk further and become healthier. They cite evidence that such experiences reduce stress and fatigue, and there is a positive effect on mood.

In the study by Beatley and Newman, 'participants showed marked decreases in depression, anger, tension, confusion and fatigue and increases in vigor' after walking in natural environments.[31] The effect also applies to environments in which animal life is present:

> Viewing birds (and listening to their calls) and watching other wildlife, at once provides mental and emotional connections, stress reduction and other biophysical benefits.[32]

Fauna, particularly bird life, 're-enchant', contribute to a 'distinctive sense of place', and reduce stress.[33] Such connections with nature also arouse curiosity and a sense of wonder.[34]

The authors identify obstacles to this communion with nature, such as busy schedules, overwork, competition from indoor activities and car dependency.[35] Computers come in for criticism:

> The time spent by children on electronic media has actually increased in recent years, boding ill for the kinds of contact with the natural world that will foster a lifelong love of and comfort and wonder provided by nature, as well as the physical exercise and activity that outside play generates.[36]

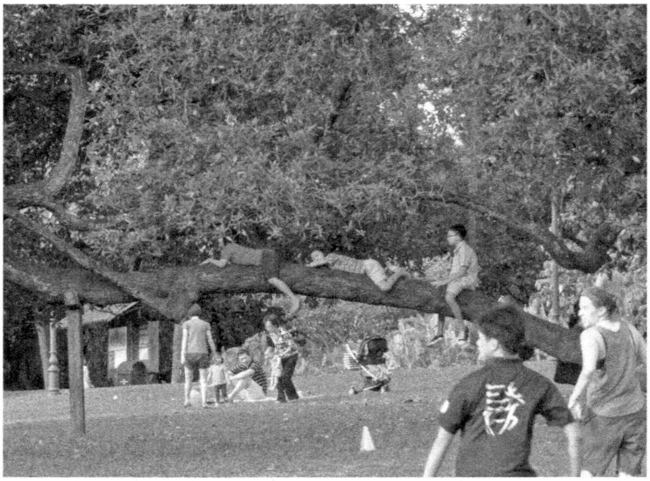

Figure 9.2 Families at play. Botanic Gardens, Singapore.
Source: author.

As I have outlined several times in this book, arguments against practices involving sophisticated technologies come easily, whatever the evidence, and despite the growing industry in mobile apps that assist communication, navigation and encourage access to outside nature.[37] It is worth reiterating the obvious, that our devices are complicit in the formation of nature. Not least, professional and amateur photography mediates awareness of nature and wildlife.[38] There is an argument that our love of animals and nature in general is already mediated by digital technologies. My message throughout is that sentiments focused on a return to unmediated nature are informed especially by the technology from which people wish to escape.

Solitude

Refuge implies solitude. One of the benefits of nature is the access to solitude it provides, whether in open spaces where solitude is available because of the distances between people, or in groves, caves and secluded woodland settings. It is common even among groups travelling in the countryside to steal moments of singular or intimate solitude.[39]

I have referred a few times to Thoreau's classic book *Walden*, which also provides an example of the natural retreat for the stressed city inhabitant.[40] Martin Heidegger also retreated to the Black Forest that provided a place where he could focus his thoughts and live the simple life away from the city.[41] As well as retreating to caves, groves and other naturally occurring refuges, people build temples, sanctuaries and shelters. Thoreau occupied a hut with a stove, as did Heidegger, and people with the means build holiday homes and baches. People also resort to a range of culturally and socially defined refuges. As we shall see, they even retreat to digital spaces as refuges.

The cultural sociologist Tia DeNora presents *asylum* as another word for refuge.[42] An asylum provides 'ontological security'. It is a space for validation, providing a sense of fit, comfort and focus. The asylum seeker withdraws from formal interaction with others, the flows of information that make incessant demands on us. You can transport yourself to a place of refuge: a special room perhaps, or some private activity, exercise, intoxicants and entertainment – your regular 'cave'. But resourceful people also convert or refurnish their environments into asylums, wherever they find themselves. DeNora generalizes the idea of refuge. Such adaptations occur where people engross themselves in conversation, dress up, participate in organized religion and participate in arts and other practical activities (choirs, bands, cookery, gardening, blogging). In other words, anything that actively engages us qualifies as a resourceful adaptation of one's circumstances into a self-made, resilient refuge.

The payoff when you either find or make asylum is respite, 'recovery of self',[43] and relief from the pressure to perform and conform. In the case of do-it-yourself (DIY) asylum creation, the asylum-maker assembles stronger resources for the next 'call to act' in her dealings with others. For DeNora, to rely on ready-made asylums carries the risk of alienation, loss of social skills, and a 'shrinking social presence'. People who make their own asylums risk a tendency to egoistic and overly assertive behaviour. But such resourceful asylum makers are in a better place than the emotional cave dwellers who retreat in isolation. DeNora draws on and develops these insights from the seminal work by the ethnographer Erving Goffman.[44] As some of the techniques for creating their own asylums people listen to music and deploy technologies such as smartphones and personal listening (and viewing) devices:

> Digitized music, coupled with miniaturization (iPods and iPads, MP3 players, smartphones), offers many more possibilities for musically inflecting and managing spaces and thus, in the process, for seeking musical asylums.[45]

At the time of writing, newspapers reported how many in the United States and elsewhere feel stressed and were in need of respite, if not asylum, from a national leadership that offers little in terms of confidence, empathy or hope. They find a kind of therapy, catharsis and place of refuge in humour.[46] Comedians and satirists have become the new leaders of hope,[47] and their YouTube clips the new refuge.

DeNora describes how we find refuge in music and other cultural forms, including while living and working in cities, but she could have been writing about nature settings. It seems therefore that natural environments are not the only providers of places to restore.

To reiterate, it is common to think that nature provides places of refuge: groves, trees, caves, patches of sun or shade, water, rocks and cover, depending on the needs of the species. Natural places (natural refuges, gardens) aid recovery and resilience to be ready for the next challenge. The anti-stress palliative proceeds as follows: to recover from the stress of the day go out into the garden or the countryside and allow yourself to engage with the intricate complexity of the natural world. If nature is not the refuge then at least there is sanctuary in its study and contemplation, as we draw inspiration from its processes of adaptation, invention and semiotic communication.

I think DeNora's insights into sanctuary expand the putative benefits of nature not only to the physical presence of nature or of our being in it, but also to reading about and studying nature, watching nature documentaries as well as engaging with its histories, issues of environment, ecology and sustainability. The acquisition of knowledge is a kind of retreat, if not therapy. Educational philosophers have said as much about learning, as edification, a building up of expertise and character.[48] Information systems and digital media (smartphones, etc.) are also tools for making refuge. They can distance us, but we can also learn through them, and our devices also deliver soundtracks for living. They are also tools that can help us discover and bring closer things that are out of reach. They filter, record and enable new forms of engagement and sociability in unlikely settings.

Nature and media

I have referred to the relationship between music, harmony and attunement in relation to nature and health. The benefits of natural environments do parallel those of certain media and art forms: 'A child in the dark, gripped with fear, comforts himself by singing under his breath.'[49] This is the opening sentence of an essay by Gilles Deleuze and Félix Guattari called '1837: Of the Refrain'. Then follows an exposition on the power of rhythm and melody to mark a refuge: 'The song is like a rough sketch of a calming and stabilizing, calm and stable, center in the heart of chaos.'[50]

It is well known that music encourages a mood, supports a sense of well-being and associates with place.[51] People also retreat into music and other listening experiences. DeNora observes that 'personal listening is also, by definition, a highly individualised solution to the problem of wellbeing'.[52] She draws on those who advocate an 'ecological perspective' on health and well-being. Once we broaden health and illness to an understanding of 'the interconnections between belief, social practice and physical embodied phenomena',[53] then it is easy to accept that a cultural intervention such as a piece of music has an influence on well-being. She writes about the placebo effect as evidence of this complex of interrelations. I am not sure she goes this far, but if you believe a piece of music is good for you then it may have just that effect, under the right conditions, much as a walk in the park is healthy if we believe it should be.

In an article 'Semiotics and the placebo effect', bioethicists Franklin G. Miller and Luana Colloca invoke Peirce's semiotic theory to explain in part the body's 'natural' capacity to heal and restore. The patient comes to the surgery or other curative setting with expectations born of a history and memory of previous conditions and encounters. Many of the beneficial effects on the patient are responses to the reassurance brought by the clinical setting, the words of the carer, the rituals of the treatment and a conditioned response from previous encounters with the treatment. Apart from its advertised causal benefits, the drug or treatment persuades the body to release its own pain and stress-relief mechanisms.[54] Here the body is a reader of signs and responds accordingly, even though the patient is unaware of the signs or the process. Such signs can emerge from the clinic, the home, a nature setting or music.

The question of why music is good for us supports further my interest in the placebo effect. Some scholars think that the mechanisms by which music affects our emotions and our well-being derive from human responses to the natural environment, as if musical sounds derive from calming and alarming sounds in our environment that signal safety or hazard, respectively.[55] This is no doubt a pertinent insight, but the concept of an interconnected whole of which biosemiotics speaks is potent as well, and accords with what geographer Jon Anderson labels a 'relational sensibility'.[56] The natural world includes us, and is made up not of independent objects as if encyclopaedia entries, but of sets of interdependent relationships: the bird on the branch in the breeze, with the insects, the sun and the rain. Add to those relationships the semiotic web that is cultural context: forests, clouds, birds in cages, melodies that sound like birdsongs, feathers as ornaments, fascinators, stories about birds, bird metaphors, bird-preservation societies, flight, aeroplanes, angels and dinosaurs. To apprehend such relationships and to see the connections is to form a bond with the natural world and the world of nature as culture.

We may suppose that technologies can jeopardize such relationships, but they can also enhance them, or make them conspicuous. People do use media that draw on and reinforce the character of the natural world. The wealth of art and architecture draws on and even defines nature themes, as do therapy practices, which again define the responses they are supposed to deliver. Meditation, relaxation and sleep aids invoke nature scenes to relieve stress and induce relaxation, such as when they encourage listeners to imagine the warm sand beneath their feet on a deserted beach, approach a clearing in a forest and smell damp pine needles, enter an exotic garden with birds and flowers, or listen to a babbling brook. Recordings designed to induce a meditative state often incorporate music. Cultural forms such as music, media and environment relate in many ways, and the relationship between the salutogenic benefits claimed of each help to reinforce the others.

Cultural conditioning

In many respects the benefit of nature is a simple message to sell, and it is easy to find supporting evidence, particularly as researchers tend to look for confirmation of a proposition they want to support, rather than the converse. We repress the obvious disadvantages of being outdoors. Human presence can harm natural environments, and there is direct harm to us: UV from the sun, bush fires, predators, perishing cold, searing heat, getting lost, being mugged or getting swept away by rivers. There are good reasons to stay indoors. In fact, *Homo sapiens* are happy to step outside when it is safe and sunny and retreat into the shelter when darkness falls, or when the weather gets unpleasant. Thoreau's hut had a stove for heating after all.

For the hunter and the hunted the edge provides cover. Species compete for the advantages provided by the edge condition, the threshold between open space and enclosure. It is the site of biodiversity. A park or garden provides a series of controlled edges. The arguments I have presented so far in favour of natural environments refer less to the untamed outdoors than the controlled and manicured world of the garden that epitomizes safety, well-being and health for the human species. I referred to gardens in Chapter 7 on play. Gardens are deeply cultural artefacts, in their design and in the stories we construct about them, as are national parks, forested areas, moors, meadows, wetlands, beaches, planes, dunes and other spaces managed for the resources they provide.

Researchers seek evidence for the health benefits of being outdoors thanks to the wealth of cultural affirmations not least supporting people's affinity with gardens: for example, the Garden of Eden, and the myth of the Primitive Hut, the first dwelling fashioned out of the trunks of trees.[57] At the same time, any experienced, perceived, reported or measured health benefits to individuals are moderated, if not determined, by cultural factors. It is tempting to say therefore that what we observe about nature is a product of cultural conditioning, as if we need to see past such predilections and prejudices, taking nature as it is. But we could also say that there is really nothing other than the world as we experience it, which is forever changing.

Natural causes

The proposition that nature settings engender health puts the matter of causality under the spotlight. Does being in nature cause us to be healthful? As discussed in Chapter 3, Sebeok argues that the function of the indexical sign speaks to a relationship more basic than causality. We think that smoke is an indexical sign of fire, and fire is the cause of the smoke. But causality does not provide the only account of an indexical relationship. Consider walking, a pastime that often takes place outdoors.[58] 'Walking cuts risk of stroke in men,' said a headline in the UK's *Evening Standard*.[59] Scarcely a day passes without official confirmation of the health benefits of walking. A person in good health exhibits signs that they exercise in the open air.

The identification of causes is an old problem. Aristotle said, '"Why does one walk?"; we say: "That one may be healthy"; and in speaking thus we think we have given the cause.'[60] Interestingly, Aristotle provides walking as an illustration of the most advanced type of causal relationship, the efficient or teleological cause. We might say walking with intention, in this case to remain healthy, but the 'causes' of any beneficial effect are difficult to identify. Is it the walking, the exercise, the psychological effect, or does it happen that people who walk a lot also eat well or have the right genetic makeup? Harking back to Chapter 7 on play we could say that to identify a cause is to conduct an investigation, to undertake detective work, compare competing hypotheses and undertake what Peirce describes as a logical process of *abduction*. To alight on a cause is the result of evidential reasoning. Aristotle knew about the multifaceted nature of causes. For Aristotle, causes are explanations of change. Whenever we ask 'why' we are appealing to a cause.

Rather than Aristotle's archaic illustrations, consider a contemporary circumstance. Imagine you are in the countryside with your pet Staffordshire terrier. You throw a ball. Why does the ball bounce? Because it is made of rubber. That is its material cause. Why does it roll across the grass? Because it is round, the formal cause. Why is it rolling now? Because I threw it, it landed and it has some extra momentum, the functional cause. Why is it moving at all? To keep the dog occupied. That is the efficient or teleological cause.

Now think of the walker. Why does a person walk? Because it is a property of nerves and muscles. That is its material cause. Why does the body progress along the path? Because the body tilts to the incline of the path. The body falls momentarily and instantly extends a leg out to stop the fall repeatedly and with alternate legs.[61] This is the formal cause. Why is the body walking now? Because it needs to get to the brow of the hill, the functional cause. Why is the body moving at all? To keep fit, healthy, sociable, engage with its animal companion and enjoy its environment. That is the efficient or teleological cause.

Martin Heidegger elaborated on Aristotle's four causes[62] concluding that all causes coalesce on the same point, which is 'to occasion'. Heidegger thinks of this as the true meaning of 'cause'. To cause is to occasion, to bring forth, to reveal, achieved sometimes in quiet contemplation, or unreflective engagement in the world, without insisting on causal explanation at all. Meditating or engaging with a nature setting sounds like

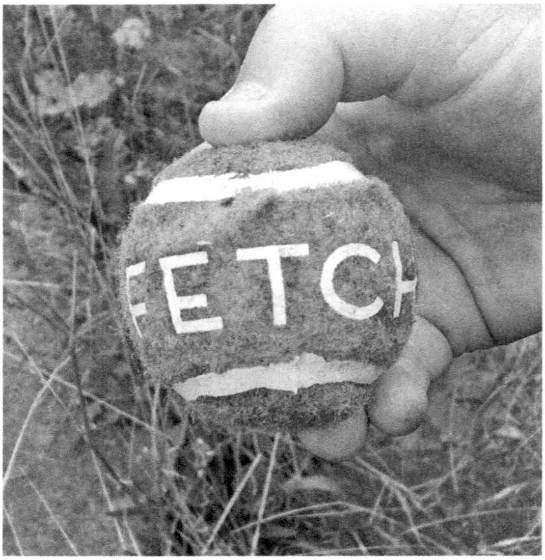

Figure 9.3 Walking the dog.
Source: author.

Kaplan's soft fascination enjoyed by nature walkers, who focus their attention on the richness of the environment – without cognitive demand. To bring this reflection further down to earth, we do not need to have a reason to walk. It is sufficient just to walk.[63] Walkers often wander and deviate, deflected to other purposes than the ones they started out with. Ambulatory practices, which connect them to their primordial grounding, simply compel them to walk. For sociologists Jean-François Augoyard and Henry Torgue walking is like talking. It is 'fluid, prone to digressions, capable of forgetting what is apparently essential and of lingering over details'.[64] As with language, the ordinariness of walking becomes a means of affirming one's presence, a 'tactic of everyday life', a 'mode of being'.

Despite such beckoning towards the unregulated, open and immediately engaging in nature, the quest for causes runs deep in the human psyche, at least in the scientific age. Apart from his theory of abduction, Peirce addresses the issue of causes, with the interesting insight that to speculate about causes is to absorb oneself in thought, to meditate in silence, and to ponder, processes that he calls 'musement'.[65] In his essay entitled 'The neglected argument for the reality of god'[66] he refers to 'Pure Play', which 'has no rules, except this very law of liberty. It bloweth where it listeth. It has no purpose, unless recreation.'[67] Musement involves wonder at the universe and 'speculation concerning its cause'.[68] By this reading, even to speculate about causes is to embark on a task without cause. Though I began this section on the theme of causes, the conclusion returns to the putative benefits of attending to nature's soft fascinations, though I maintain that such benefits accrue by many other means as well.

A simpler world

An influential 1991 academic article by environmental psychologist Roger Ulrich and his team reiterates the putative benefits of being outdoors.[69] They support the human tendency to enjoy natural environments and to find them restorative, pleasurable and mood altering (in a positive way). They maintain that nature settings are simpler than complicated, frenetic and stress-inducing urban environments. The researchers offer support for what I outlined in Chapter 1 as the cognitive benefits of moving from a complicated environment to a simpler one, the broad vista that encourages an aha moment. Of course, what constitutes a simple environment depends on your experience and point of view. It is really about familiarity. A visitor must spend more time thinking things through when in an unfamiliar setting. I assume that unfamiliar (urban) environments require more cognitive effort, and so do not provide respite and time for recovery.

But Ulrich et al. also refer briefly to arguments in favour of nature settings based on the idea of 'cultural conditioning'. They say that 'contemporary Western cultures tend to condition their inhabitants to revere nature and dislike cities'.[70] We revere the natural against the technical. They do not articulate reasons in their article for how such cultural conditioning might take hold. But an appeal to 'culture' already alerts us to the power of language, not least the establishment of oppositions such as nature/city and natural/artificial. The researchers do not discuss this, but those inclined to Jacques Derrida's philosophical challenge to entrenched cultural oppositions would assert that every appeal we make to the existence of nature in the raw is already imbued with artifice, that is, technologies. As I have already indicated, apart from the language we use to describe nature, we see landscapes through the lens of so many paintings, photographs and works of literature, mediatized, enhanced, promoted and filtered. When we are in nature's settings we wear appropriate clothing, hiking boots and carry guidebooks in nylon backpacks and carry smartphones.

Whether or not mobile technologies make life in the countryside simpler, and therefore help sustain its restorative benefits, they do offer familiarity. They provide the opportunity for sociability. They offer a further potential for walking with someone. There is comfort and safety in numbers, virtual or otherwise. Smartphones also have the potential to enhance curiosity and fascination by virtue of all that information at our fingertips. They are part of what it is to occupy a world both familiar and strange, which is sociable, linguistically rich and packed with information.

Loss of nature

From the foregoing discussion, it appears to me that nature comes to the fore most forcefully as people consider what they have lost in this highly technologized world. I alluded to this loss a few times, in the Introduction and when discussing the

Anthropocene. Nature is a collection of lost benefits. Memories of childhood and lost innocence seem important if a writer is to participate in the experience of nature. The influential book *Last Child in the Woods* by Richard Louv expresses regret over our loss of contact with nature, and provides a rallying cry to get back in touch. Louv uses the phrase 'nature-deficiency disorder' to describe 'the human costs of alienation from nature, among them: diminished use of the senses, attention difficulties, and higher rates of physical and emotional illnesses'.[71] He thinks that the 'disorder can be detected in individuals, families, and communities'.[72] It is difficult to deny the value of nature, and being in it, especially when aligned so strongly with childhood and its innocence. As I have indicated, it is also easy in the same breath to disparage the effects of modern technology: 'Experience, including physical risk, is narrowing to about the size of a cathode ray tube, or flat panel if you prefer.'[73] Louv thinks that due to networked computers we are growing 'more separate from nature' and each other.

Such sentiments find ready resonance among electronically connected contemporary adults who recall a time when things were different. It is worth recapping on what people claim we are losing in the digital age. First are the claims that orbit around the idea of attunement, connection and participation in the biosemiotic sphere. Second, I have alluded to the possibility of solitude. Most accounts of a nature experience among the romantics were of the lone wanderer savouring the wonders of the natural world. Perhaps solitude brings benefits. It is easier to achieve in the vast expanse of the outdoors. Critics think of potential solitude and isolation as the least desirable aspects of life online. A third loss if we are distant from nature is company. Camping, hillwalking picnics, bush craft and white-water rafting are group activities. Projects in the outdoors typically require some level of cooperation. Perhaps it is intense project-based sociability that yields benefits. The fourth loss relates to altruism and its rewards. Whatever the benefits to individuals and the human species, the world outside of human artifice needs to be sustained and respected. To experience nature is to engage with this bigger project. What some call 'altruism' loops back to tangible and emotional benefit. Fifth, as discussed in the case of walking earlier in this chapter, exertion carries health benefits, which we lose when away from nature. Nature experiences afford opportunities to traverse distances, work against physical resistance, run about, breathe deeper and become fitter. Sixth is the loss of knowledge. There is a lot to know in nature, and people benefit from being able to recall, identify, classify and explain it, 'the power to guess nature's ways'.[74] It is a rich arena for the exercise of shared practical wisdom. Seventh is the loss of challenge. Nature experiences afford some degree of challenge in requiring resourcefulness in the face of problems and dangers. You can fall from a cliff, get stung by a scorpion or get lost in cloud on a Munro. The occasional danger nature throws at us stimulates even the most risk-averse.

Eight, human beings are losing contrast: moving into and out of safety and danger, light and shade, the simple and the complicated, the familiar and the other – leaving the safety of home for the wild unknown, and returning with some new insight or point of view, only to venture out again in the constant transformation that is life. Perhaps there is something about the 'frontier spirit' here. This list can be extended further. Perhaps we have lost the ability to do things with our hands (climb, pick up sticks, caress a rock), play, run, interact, give scope to the full range of the senses, meditate, let be, exercise our hardwired biological affinities, participate in the web of life, fill our lungs with fresh air, absorb UV light, or connect with as-yet unmeasurable natural fields and forces (to be explored in the next chapter).

Perhaps we miss doing what just feels good to do. Stamp collectors, guitar players, sociable drinkers and bloggers move into a beneficial frame of being when doing what they like doing. If you like the forest, then you anticipate and experience some benefit from being in a forest. But some measure of anxiety follows if that privilege is neglected, forgotten or denied. Perhaps natural environments afford these benefits bundled together, in variable intensity, mostly delivered at low cost, without special equipment or the need of a checklist.

It does appear as though scholars and enthusiasts for computers, digital media and devices can advance claims like those offered in favour of nature settings. Artefacts, buildings and experiences induced by artificial means, are in the company of music, watching movies, hanging about indoors, indoor exercise and looking at abstract pictures and patterns. Commentary on urban living in general can draw on similar claims. That is not to valorize computers, networks, mobile devices or the built environment (or to ignore their shortcomings). On the one hand, it indicates the challenge: What does nature afford that the world of technology does not? On the other hand, it offers a resolution. One solution to the *aporia* is simply to accept that the nature–technology dichotomy does not serve us well in the cause of health and well-being.

I have referred to the well-being of the human in a nature setting. One of the side benefits of such engagement goes towards nature. The more comfortable people are with nature settings, the more likely they are to value, protect and preserve them. It appears that the greatest predictor of care and respect for natural environments among adults is the exposure to nature that they had as children.[75] But other technological resources are also of value, including classroom study of the natural world, including mathematics, writing and history. Neither does divorce from and denigration of such skills and the technologies that support them serve nature well.

Resilience

As I have discussed, balance is one of the master metaphors of health, life and of nature. Living with nature makes us healthy and resilient. The seminal book by George Lakoff and Mark Johnson, *Metaphors We Live By*, identifies balance as a ubiquitous concept that they trace back to the human body. After all, from an early age, the human animal becomes aware of the need to stand and to balance on just two spindly limbs. The body has a symmetry that engenders the idea of a left and a right – in balance. Such symmetry translates to the way we reason and speak – weighing up the options, deciding on balance to go for a walk or watch the television news, and other more serious ethical challenges. The famous icon of justice is a female figure, blindfolded and holding a pair of weighing scales for comparing the weights of two commodities.

As I have shown, traditionally, many scholars thought of health and well-being as matters of balance, for example, the balance of the bodily humours. In his book *Attunement*, and in which he advocates for the theme of balance, Alberto Perez-Gomez says of ancient classical architecture and its successors that

> the role of architecture, particularly the city's orientation and its buildings' properly proportioned configuration, was to mediate between man and nature, the living cosmos, and thereby contribute fundamentally to the maintaining of such balance, allowing humans to live harmonious lives.[76]

It seems that for Perez-Gomez our modern, highly engineered and digitalized age loses this balance. Not only the body, but nature and our relationship with it becomes the model of a balanced existence, much desired and highly valued. Balance, harmony and equilibrium come across as necessities and virtues in the worlds of architecture, environment, physical and psychological health and nature.

It is strange therefore that balance is not the keenest metaphor environmental scholars deploy when discussing nature. Ecosystems are often characterized by wide fluctuations in populations across species as each competes for dominance. Such is the Darwinian struggle for survival. Then there are catastrophes, even independent of human intervention, where fires, floods, eruptions, climatic change and competition destroy whole populations. That is nature 'raw in tooth and claw'.[77] We can agree with the critical theorist Herbert Marcuse: 'The world of nature is a world of oppression, cruelty and pain, as is the human world; like the latter, it awaits its liberation.'[78]

The balance narrative offers a highly selective view of natural systems, equated with a condition of stasis, stability or a steady movement towards 'improvement', such as a recovery of indigenous species, the retreat of invasive organisms, a restoration of biodiversity or some condition of apparent cooperation between human habitation and that of other kinds. For the naive it points to the eradication of parasites, pests and diseases.[79] The balance narrative applied to architecture paints a utopian picture of an impossibly balanced universe where 'the wolf will dwell with the lamb',[80] the temple is at the centre of the town, the yearly cycle divides neatly into 360 days, and the earth's axis is perpendicular to its path of transit in a circle around the sun, accompanied by a lunar orbit in perfect synchrony. The rhetoric of balance points to a static universe, not entirely beneficial to the human organism in the natural world. Organic life by most accounts is a product of eccentricity, deviation and the discrepancies that arise when things do not align – and in their lack of balance, deliver tides, seasons, regions and margins, enabling change, growth, innovation and organic diversity.[81]

In an influential article on ecological systems, Crawford Holling emphasizes the metaphor of *resilience*. Contrary to the idea of stability, there are highly unstable ecological systems that demonstrate 'an enormous resilience',[82] where 'instability, in the sense of large fluctuations, may introduce a resilience and a capacity to persist'.[83] Such observations reinforce Friedrich Nietzsche's famous aphorism: 'From the military school of life. – What does not kill me makes me stronger.'[84]

Irrespective of benefits to psychological or physical health, such tendencies towards resilience accord with the instinct among many of us to seek out diversity in our experience of places, social situations, people and life in general. Some of us, for some of the time at least enjoy bucolic landscapes, well-proportioned buildings, meaningful places and profoundly artistic experiences even more when they appear against a backdrop of the mundane, tedious, risky or downright ugly. Such variety can in turn transform the mundane into the intriguing, zany, challenging and a kind of sublime.

In this foray into nature as refuge, I considered attunement, healthy environments, solitude, music and other media forms, cultural considerations, causality and loss. Once we accept the semiotic character of the natural life world and our place in it, then we must also admit digital technologies, which are after all contributors to the wider semiotic sphere of a kind of 'relational sensibility'.[85] People use devices, including smartphones with touch screens and headphones to adjust their responses to environments. Once we broaden health and well-being to an understanding of 'the interconnections between belief, social practice and physical embodied phenomena',[86] then it is possible to accept that what we choose to think of as nature is amid a range of factors that impinge on well-being, balance, resilience and decay.

CHAPTER 10
NUMINOUS NATURE

Who can deny that there are places whose 'numinous nature' is best sensed

> in the peace and tranquility of the atmosphere that surrounds them, simply by sitting still and quietly letting the energy and power of the place slowly seep over you.[1]

This passage is from a book on ley lines by Christopher Street.[2] A numinous place is occupied by a divinity, or it can simply 'evoke a heightened sense of the mystical or sublime'[3] according to the *OED*. The mapping of lines connecting ancient landmarks in and around London informs Street's account. He refers to the seminal text on ley lines by the photographer and amateur geographer Alfred Watkins (1855–1935). Ley lines are the lines of sight used by prehistoric surveyors to position and align monuments with significant land features. It seems that Watkins saw no need to explain the phenomenon in terms of hidden forces. He did, though, add raw material to an industry of speculation on the mysterious forces of nature.

Of those ancients entrusted with the responsibility of mapping and siting he asks whether 'they make their craft a mystery to others as ages rolled by'.[4] Perhaps the ancient planners were the priestly Druids as the Romans thought, and as the ley declined, it degenerated 'into the witches of the middle ages'[5] who rode on broomsticks: 'They (in imagination) flew over the Broomy Hills and the Brom-leys.'[6] He adds: 'It may be that the ancient sighting methods were condemned as sorcery by the early Christian missionaries.'[7] He speculates further: Were the ancient planners 'the laity or lay-men of Beowulf?'[8] Such unanswerable questions bind mystery with magic.

There are other ways that people think we connect mysteriously with nature and the character of a place, among them astrology, geomancy, divination, feng shui, morphic fields, alchemy, hermetic philosophy and shamanism, each with different attachments to evidence and authority, and variously supported in particular cultural contexts.[9] As a further example of the significance attached to coincidental arrangements, enthusiasts are able to generate maps showing Chartres Cathedral at the centre of a great circle that has Rome and Sintra (Portugal) on the circumference.[10] Equidistant from each of those two cities and on the same circle lies Rosslyn Chapel just outside Edinburgh.[11] Certainly Rosslyn attracts interest due to the unusual shape of the architecture. The stone masonry detailing is out of scale with the usual Norman or Gothic stonework, and the chapel has apparent associations with the Knights Templar. The site features in Dan Brown's mystery novel and the film *The Da Vinci Code* (2006).

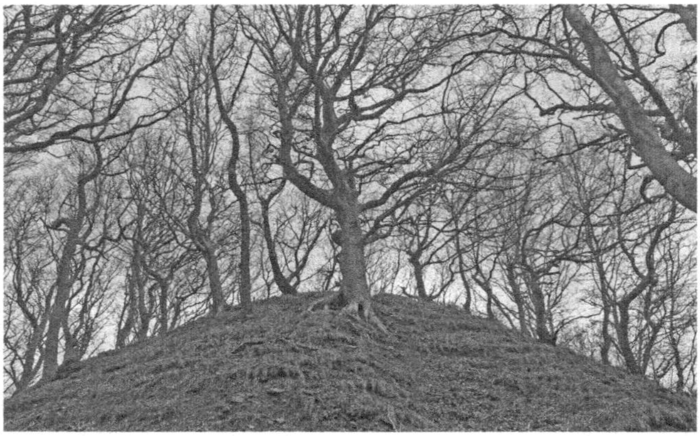

Figure 10.1 Numinous landscape. Site of Roman fort, High Bradfield, South Yorkshire. *Source*: author.

Can people connect directly with the 'energy and power' of an environment? Some people think of the mood of a place in these terms.[12] There are haunted places, and buildings steeped in atmosphere and memories. For some people this connection is something other than poetic analogy and metaphor. It is even more than symbolism. It is as if certain places emit a vibration, a field, and provide a direct organic or parapsychological connection between people and places. The myth of numinous places enlivens, enchants and provokes fascinating stories and mysteries about a place. Their plausibility fails though where they rely on scientific, empirical and material causal narratives. My purpose in this chapter is to demonstrate that such exotic narratives sidestep the more prosaic but pragmatic factors that contribute to our sensitivity to place, such as interpretation, the semiotic field, cultural context and metaphor. In case a world so defined appears boring, I conclude with the proposition that to apprehend the mundanity of a place can enrich our experience.

Nature's apothecary

The so-called natural environments (the outdoors, hills, forests, meadows, parks, gardens, lakes, deserts) provide a good test case for the proposition that there is something more to nature than science has yet revealed. For one thing, some people think there is an organic connection between people and natural environments – a set of connections disrupted by artificial intrusions, such as roads, factories, power stations, wind farms and a society under the sway of smartphones, screens and digital networks. After all, humans and environments co-evolved. There must be many kinds of connections yet undetected by science. Perhaps people have a sensory attunement to aspects of the environment of which they are unaware, or to which they cannot give clear expression. All they can say is that they have a certain feeling about a place, or that it exerts an influence on them in ways difficult to identify. A phenomenologist would say that such feelings are due to our embodied being in the world. The related semiotic position as put forward in this book is that we are connected by sign systems. It is not simply the case that we currently lack the science to pin down the requisite cause and effect.

Perhaps we are influenced by signals from the environment that are out of the range of our conscious perceptions. Semiotic theorists acknowledge that signs can be detected and acted upon without conscious attention. Peirce alludes to the condition of a friend who lost his sense of hearing, after which the friend would enjoy 'listening' to a piano. When standing next to a piano the friend claimed that he could 'feel the music' all over his body. Perhaps the friend had some hearing after all. But the friend asserted that this is not a new sense: 'Now that my hearing is gone I can recognize that I always possessed this mode of consciousness, which I formerly, with other people, mistook for hearing.'[13] Peirce draws a lesson by analogy about consciousness: 'In the same manner, when the carnal consciousness passes away to death, we shall at once perceive that we have had all along a lively spiritual consciousness which we have been confusing with something different.'[14] Peirce also alludes to a 'social consciousness, by which a man's spirit is embodied in others'.[15] He expands this proposition as a truth 'which is embodied in the universe as a whole'.[16] Peirce labels such a unity 'synechism',[17] a position that starts from the presumption that 'everything is continuous'.[18] Here he counteracts the dualisms that have arisen in European philosophy, such as the distinction between matter and ideas, body and mind, and in making his assertions he shows his allegiance to various traditions of *monism*, the view that the universe is one whole.[19] This is a view consistent with his position on semiotics as communication via signs within the natural world.

That nature harbours organic connections of which we are yet unaware is a view familiar to contemporary movie watchers and enthusiasts of fantasy and science fiction. To be more precise I should say that fantasy and science fiction feed off some primitive urge in all of us to identify connections in nature. For example, the film *Avatar* (2009) features the 'Tree of Souls', where the planet moon's life source reaches out to everything through roots and tendrils, materially connecting organisms together, at least on that particular moon. There is also 'The Force' in the *Star Wars* (1977–2015) series. Those individuals and families in whose cell structure resides a high concentration of microscopic 'midi-chlorians', have access to the Force and its powers, especially if they train as Jedi knights. Theologian and Heidegger scholar, John Caputo, thinks that the 'religion' of *Star Wars* sidesteps long-standing debates about faith and reason, and the natural and the supernatural: 'The gifts that the Jedi masters enjoy has a perfectly plausible scientific basis',[20] which is not to say the storytelling delivers that structure consistently or without controversy among the many fans and critics of the franchise. For Caputo, the spiritual, paranormal and magical in such fantasies acquiesce to a kind of scientific causality. From the point of view of this chapter though, it is interesting that such explanations are grounded in the organic, the stuff of nature. Life still holds a mystery for us, and such narratives are more compelling to the modern sensibility when they appeal to organic life rather than abstract concepts from physics and geometry, let alone a spiritual realm appropriated by faith. The Force of *Star Wars* is a term in physics (force equals mass times acceleration). As a contemporary religious idea and as a term in physics, 'force' becomes shorthand for pervasive biological processes, albeit not yet fully understood. And such organicism has a spatial aspect. 'You must feel the Force around you. Here, between you … me … the tree … the rock … everywhere!'[21] said Yoda in *The Empire Strikes Back*, and 'the force is strong in this place', say several fan-fiction spinoffs.[22] The force and 'midi-chlorians' find a place in popular fiction and science fiction, albeit with mythic roots.

There are other theories about the invisible effects of place that claim a grounding in science. One such approach is to draw on quantum theory to posit the existence of undetectable fields of action known as 'morphic fields'. Re-enchantment seeks to hypothesize, investigate and explain in terms familiar to our twenty-first-century scientific sensibility what physical science is not able to account for readily, or even to detect. That is my interpretation of the project of Rupert Sheldrake and others. Here is one of their propositions about 'morphic resonance':

> According to the hypothesis of morphic resonance, human beings draw upon a collective memory: something learned by people in one place should subsequently become easier for others to learn all over the world.[23]

Unlike magic, this kind of speculation posits cause-and-effect explanations in terms of quantum physics or some other (difficult) field theory. A famous example is the observation that a species of birds (blue tits) in one region of England discovered how to remove the caps from milk bottles and drink the cream. The habit spread throughout the country (among blue tits), even though there was no direct interaction between colonies of birds. The strong claim of morphic resonance theory is that birds of a species, or the elements of any other 'self-organizing system', are interconnected by a morphic field that aids learning. It is a form of action at a distance.

I remember the days where we had to coach students on how to hold and click a computer mouse correctly. Morphic-field theorists might say that thanks to morphic resonance, once a few people have been taught the skill, everyone else can do it instinctively. As with other enchanted pursuits, such explanations diminish the complex social and spatial aspects of cultural transmission – how communities learn. Were the morphic effect not so elusive we would expend less energy training and educating people, or researching the ways human practices are acquired and transmitted. Classes and the bill for education and R&D could be much smaller if we understood how to tap into morphic fields – if they existed.

Then there are some causal stories asserting our innate connection with earth, beyond the connections posed in Chapter 6. The radical cultural theories of Walter Benjamin lament the loss of aura evident in automation and the endless reproduction of the visual image. He posits several definitions of aura, not least it is a sonic (aural) display, the glow seen along the ridge of a distant hill and a gentle movement of air. The *OED* also refers to an aura as an 'electrical discharge', or 'the current of air caused by the discharge of electricity from a sharp point'.[24] Electricity looms large in numinous connections to the earth. An intriguing article by biologist Gaétan Chevalier and his team suggests that indoor living means we are losing contact with the earth, that is, that our bodies are insulated from the ground with which the current of electrons into and out of our bodies would otherwise flow.[25] They suggest that 'this disconnect may be a major contributor to physiological dysfunction and wellness.'[26] Contact with natural environments (particularly the ground) allows us to 'equilibrate with the electrical potential of the Earth'.[27] This proposition is not only about health, but also about the removal of anxiety, depression and irritability. According to some of the papers cited by Chevalier et al., the cure for an imbalance of electrical charge in the human body is to walk around barefoot or at least to avoid synthetic rubber shoes. You can also sleep on a special mattress grounded to the earth through pipes in your home, in the same way that electrical sockets are earthed. Certainly, walking barefoot has become a sign of reconnecting with our primal roots, reconnecting with our autochthonous selves, but this theory posits the flow of electricity as the instrument of that reunion with the earth.

Fantasy and computer games expand on such causally imbued enchantment. Fantasy video games instrumentalize special marks, stones, plants and artefacts, as if such objects function as autonomous causal agents able to influence events, for example, the ring that has awesome power (*Lord of the Rings, Green Lantern*). It is easy to simulate such cause-and-effect relationships in computer games. In the game *Myst IV*, in the 'Age of Serenia' people's memories float around like spores from some kind of plant, spirit sprites hover over water and fire rocks, and the inhabitants find out about the future when they dream. There is something gratifying in fantasizing about such organic causal connections (plant spores). Such stories sustain the mystery inherited from the wonders of nature. At the same time, such games present explanations, or potential explanations, that satisfy our desire to acknowledge the authority of science, in this case via pseudo science, or a science as imagined.

Think of a time before modern scientific inquiry and digital technology. Shakespeare's play *A Midsummer Night's Dream* and its performance on stage and film captures something of the 'magic' of nature, with its fairies, love potions, mystery and transformation. The play reflects the belief of the times that forests and woodlands harbour curative powers. There is little in Shakespeare that extolls scientific causality, technology, or even the beautiful, sacred and sublime in nature. The latter came much later with the romantics and in contemporary performances and adaptations.[28] Though the events of *A Midsummer Night's Dream* may have been remarkable to Tudor audiences, the nature setting was neither exotic nor mysterious. It had a familiar ring. The play echoed common carnival and fantasy themes, but according to a 1959 paper by Lou Agnes Reynolds and Paul Sawyer, journeys into nature (mostly forests) were a means of gaining access to the healing properties of plants and other folk remedies. Common folk did not need to be specialists to know about the medicinal properties of plants, insects and animals.

The play makes obvious reference to potions derived from purple and white flowers used to confound and cure the love relations among fairies and mortals. According to the researchers, 'Shakespeare seems to have taken this familiar folklore concerning the magic power of plants as a meeting point between the supernatural and the natural worlds.'[29] This interpretation of the play is useful as an indicator of the strong cultural legacy associated with natural settings as a source of healing. Not many of us now forage for herbs or consort with fairies, but as indicated in the previous chapter, we still think a walk among plants and animals has therapeutic value, and such natural commonplaces can restore, if not enchant.

Digital pagans

In the early days of the world wide web in the 1990s I reported on the overt reference to magic that found enthusiastic expression in this new medium sustained in part by the 'magic' of the computer.[30] Computer graphics images became icons, talismans and magical objects. Photorealistic computer graphics provided a plastic medium for the exploration of the bizarre, the mysterious and the surreal along romantic medieval pictorial themes. In fact, the fetish with the depiction, collection and display of vast catalogues of curious objects seemed to sustain a thread of magical irrationalism. Contrary to the association computing commonly held with encyclopaedism, analysis, increasing bureaucratization, Enlightenment and objectivity, here we had the burgeoning of a medium for the release of sentiment, subjectivity, mystery and magic.

Kevin Robins, a critic at the time of cyberspace rhetoric, highlighted the uncanny tension between imagination and reality:

> In the virtual world, it is suggested, we shall receive all the gratifications that we are entitled to, but have been deprived of; in this world, we can reclaim the (infantile) illusion of magical creative power.[31]

Around the same time, Marcus Novak suggested that cyberspace is a meeting of the objective and the subjective, though it represents a triumph of the subjective. It is a 'habitat for the imagination':

> Cyberspace is the place where conscious dreaming meets subconscious dreaming, a landscape of rational magic, of mystical reason, the locus and triumph of poetry over poverty, of 'it-can-be-so' over 'it-should-be-so'.[32]

Magic is clearly conjoined with imagination here. I will return to imagination towards the end of the chapter.

In my summary of these critical analysts of magic and mystery I referenced the film *The Wizard of Oz* (1939) where it turns out that the wizard's gifts of intelligence, courage and compassion reside in trivial symbols (a diploma, a medal, a heart-shaped clock) and the wizard's hollow valedictories that expose the foibles in the relationships between the main characters, who, after all represent the combining of human (Dorothy), animal (lion), vegetable (straw man) and mineral (tin man). This interpretation is heightened when we appreciate that Frank Baum (1856–1919), the author of the book on which the MGM film is based,[33] was a Theosophist. In its attempt to collapse the distinction between magic and science, Theosophy trades in symbols. According to commentator Susan Wolstenholme, power in Oz comes from natural forces, like tornadoes, but sometimes it resides in manufactured items (symbols) such as slippers, houses, medals and clocks, which share in the magic that is resident in nature.[34] Video games also acknowledge such an impetus, in which players collect and deploy talismans to advance the game play. Magic draws on signs and wonders, and as such recruits the play of semiotics.

Among the references to alternative rationalities online we then found variants of a new techno-paganism moving into prominence. Pantheism is one of several systems of thought and belief that aligns with a sublime respect for and even adoration of nature. Pantheism affirms 'the animating spirit of nature',[35] and that 'God is the unified totality of all things'.[36] Pantheism is in the company of contemporary polytheism and paganism. It is also present as people talk about digital technology. Nature reigns supreme, but digital networks follow in its train.

In his book *The Digital God*,[37] psychologist William Indick highlights various transformations in religious sensibility brought about by digital culture, in particular the multisensory modes of interaction provided by networked computers:

> The personalization, perceptualization, and sensualization of the experience of God may even lead a reversal back to premonotheistic spiritualities – polytheism, pantheism, paganism – resurgences of which have already been observed in Europe and America.[38]

This is a claim about the freedom of individuals to find their own spiritual home, aided by the liberating and democratizing power of the internet. But the language elides readily into the pantheistic tropes of uniting, melding and connecting: 'As digital media, cognition, and perception continue to meld into one integrated system, the doors of perception will open wide.'[39] Simulation plays a part: 'The virtual reality of multisensory digital simulation will give us the power to create our own spiritual perception, as vivid as any dream, and as visceral as any real world sensory experience.'[40]

Sharing and communion also come into play: 'The power of the internet will allow us to share our spiritual images, our dreams of God, with everyone around the world, in a truly universal church, experienced as a communal dream.'[41] This digitally enhanced revival de-privileges conventional organized religion that claims little of its authority from scientific causality. Indick stops short of attributing divine connectivity to electronic networks and the internet of all things, but he does endorse a transition to a blissful state of transcendence:

> In due time, we will devise simulators that will bypass our conscious mode of perception, tapping directly into our unconscious awareness, the dark hidden space within, in which spiritual perception is not inhibited, and is ready to be retrieved and enhanced.[42]

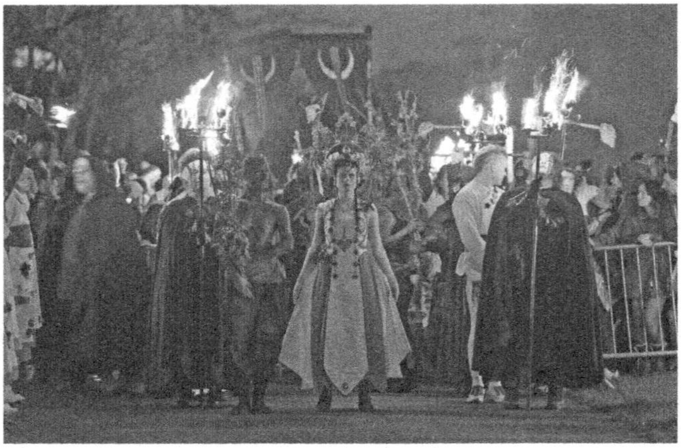

Figure 10.2 Edinburgh Beltane Fire Festival: annual interpretation of a pagan festival welcoming the May Queen and the start of summer.
Source: Getty Images: Roberto Ricciuti.

He adds, 'The digitally fueled inward journey into the kingdom of mind will at last reveal the truth of Jesus' proverb: "The kingdom of God is within".'[43] Such ambiguous sentiments provide evidence of the subjugation of enchantment to technology. Our spiritual lives are after all subsumed within ubiquitous networks.

To this end, some groups even call themselves 'technopagans', tapping into the power and opacity of computer code and hardware. According to an article in *Wired* as long ago as 1995, 'a startling number of Pagans work and play in technical fields, as sysops, computer programmers, and network engineers.'[44] Now smartphones you can speak commands to, and ubiquitous networked communications, lure us into a modern Harry Potter world of wizards versus muggles.

Such narratives are familiar and accord with other ambiguous alignments of the digital world with philosophy and religion via the singularity, super-sensible hive minds, mind-melds and other techno-utopian dreams. From my point of view, digital pantheism provides a further example, if we needed it, that the digital world and the world of nature are not so far apart – at least in the way some people, for some of the time, talk about them. The natural and the digital are united by magic.

The magic circle

Play and magic align. The historian and theorist of play Johan Huizinga (1872–1945) drew attention to the 'magical' aspects of the spaces in which people play. A place of play is consecrated and set apart for play. It is where rules different to those exercised in day-to-day living have full scope:

> The arena, the card-table, the magic circle, the temple, the stage, the screen, the tennis court, the court of justice, etc., are all in form and function play-grounds, i.e. forbidden spots, isolated, hedged round, hallowed, within which special rules obtain.[45]

Game theorist Jesper Juul provides a suitably ordinary example of where the magic circle pertains. Most people think it is impolite to snatch the salt from the table as soon as you see one of the other dinner guests make a reach for it. Yet, later in the evening while playing a card or board game, the same diners will grab at a pile of cards in the card game snap, deceive each other in poker or charge uncharitable rates to bankrupt poorer players who land on their property on the Monopoly board. What is unacceptable in polite social interaction becomes acceptable in game play.[46] Video games offer more extreme examples: inflicting violence on people and property or shooting people dead in a video game.

In fact, magic circles form around fiction and film. Authors are exempt from the consequences of the virtual crimes they commit through their characters. What would happen to the business of storytelling if courts prosecuted JK Rowling for the misdeeds of Voldemort or Dolores Umbridge! An author will not be prosecuted for killing off her characters or encouraging her characters to transgress.[47] Juul argues that the boundary of this magic circle is not fixed, especially when we think of 'meta-games'. Sometimes players want to lose because the social situation presents losing as a viable option, as when playing with a small child. A player may also make a bad move deliberately to keep the game interesting, or they lose interest and just want the game to be over. I could also add that the boundary is blurred as we think of various meta-game tactics. In the case of video games, players are known to save the game state in order to reset after failing in a high risk move. The film *Edge of Tomorrow* (2014) with the tagline 'live, die, repeat' plays on this theme of meta-game repetition, as the lead character conquers invading aliens by 'resetting' each time they kill him. Through repeated iterations he can improve his fighting skills. *Groundhog Day* (1993) and *Source Code* (2011) play with similar themes of reset and repetition.[48]

Whether in the game or at the meta-game level, we players and audiences have become accustomed to time travel and games that exhibit temporal paradox. Some games also play with spatial paradox within the game's magic circle. Audiences and gamers accept systems and devices that transport the traveller from one location to another without having to negotiate the space between. The transporter in *Star Trek* (1966–to date) is an obvious example. In the multiuser role-playing environment *Second Life*, players can position portals in key positions that transport the player from one location to another. This is a trivial operation in the Cartesian space of 3D games. The game simply assigns new coordinates to the players' first-person game positions, and that of their avatars.

Games that present impossible operations within spaces with rules contrary to nature include *Portal* (2007), which reconstructs the laws of Euclidean geometry and physics. In this game, the players aim their portal guns at walls, floors, ceilings and platforms to create entry and exit holes called 'portals'. If your avatar jumps through the entry hole in the floor, then you may arrive in the same space having re-entered through the wall. The game involves learning what you can and cannot do in this alternative universe's 'magic circle' to progress through successive levels of difficulty and reach the goal of the game.

Artists and architects have long worked with paradoxical spaces. In tromp l'oeil (trick of the eye), the artist paints some objects, or extensions to a room, as if we are looking at something in three dimensions. Architects in the Renaissance would employ fresco painters to provide an illusion of depth to a space, as if the dome reaches higher, the nave of the church extends beyond the altar and columns and pilasters protrude from a flat surface. Such anamorphosis includes distorted images that appear coherent when reflected in a cone or cylinder, or from a particular angle. There are many examples online of 3D pavement art drawn to give the impression that there is a gaping hole in the pavement, or a person standing on the pavement is balanced on a precarious ledge above a waterfall or on an island in a lake. The illusion works best if viewed from a single optimal eye position, and is best appropriated via photography and circulated online.

The graphic artist Maurits Cornelis Escher (1898–1972) produced many enigmatic woodcuts and lithographs based on tessellation patterns and spatial incongruities, many involving elements from nature. One of his most arresting images presents water flowing

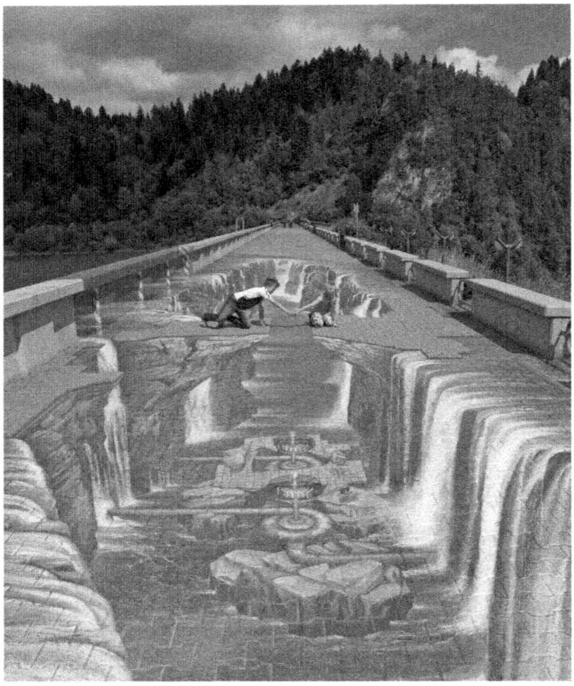

Figure 10.3 Three-dimensional illusion graffiti painted on the dam on Dunajec river on 1 July 2013 in Niedzica, Poland.
Source: Mariusz Switulski/Shutterstock.com.

along a channel. The channel recedes away from the viewer but at the end of the channel the water appears to be at a higher level than it was at its starting position. There the water tips over the open end of the channel and drives a water wheel at the lower level and continues its journey along the channel in a looped sequence ad infinitum. The presence of the water wheel suggests that this is a perpetual motion machine. The drawing is possible due to the peculiarities of isometric projection. On YouTube there are several examples of people trying to build Escher's water device out of wood in three dimensions.[49] The video game *Monument valley* that can be played on a smartphone or a tablet computer presents similar spatial anomalies via a range of isometric 3D image renders and animations.[50] Once we get used to them we move adeptly in and out of magic circles.

The ability to cause action over a distance is one of the hallmarks of enchantment, expanding the magic circle into the world of everyday products. I can adjust the thermostat on my home heating from my smartphone while in another city, and control the selection, pause and replay of movies on my television screen from my smartphone, further amplifying the magical illusion of action at a distance.[51] Advertisers feed magic into contemporary product branding. According to art critic Virginia Heffernan, 'Magic is a word that Apple vigorously embraced.'[52] She adds:

> The iPad was introduced as a 'magical and revolutionary device'. And *magic* is a crucial term of art in computer programming. Computer code is considered magic when it seems simple but accomplishes complex operations. The Internet is paradigmatic magic.[53]

This to me speaks of magic as a richly cultural meme, a major source of the power and currency of digital technologies, something it shares with numinous nature.

Nature's symbols

I have discussed symbols as one of Peirce's sign categories. Symbols are signs that link arbitrarily to their referent, as in the case of symbols in a mathematical formula, logical calculus or a computer program. Most nouns are symbols, evident from their interchangeability across different languages, and so are crests, brand labels, swords, chairs and almost any physical object. Peirce discusses symbols at length: 'The symbol is connected with its object by virtue of the idea of the symbol-using mind, without which no such connection would exist.'[54] Symbols do not gain legitimacy by being like the thing they are referring to. That is an icon. Nor are they attached to their referent causally, as lightning is a sign of a storm (as an index). Convention makes the symbol. Language communities adopt particular linguistic conventions in processing their symbols, as do elite groups of coders who know what the symbols mean in their particular computer programs.

But Peirce fails to address another important category of symbol structure that further unites nature and artifice. There are symbols that participate in important meaning structures, where one symbol cannot easily be substituted for another. Symbols derived from nature hold a privileged position in the symbolic landscape, and even assist in defining the natural. At least, symbol and referent are closely coupled. Many people know what it means to say that a tree symbolizes life, growth and strength. In the Garden of Eden, the tree is a symbol of knowledge, and is a bearer of tempting fruit, the consumption of which strips its consumer of innocence. The tree as symbol here differs from the tree as icon or index.

By their specificity, symbols bind communities together and to the landscapes they inhabit. For example, the historian of religion Mircea Eliade (1907–86) draws attention to the image of the *centre* in mythic traditions. The idea of *the centre* draws on symbols of the pillar, ladder, gateway, navel, mountain and tree, through which humans and gods communicate, linking the earthly cosmos to the realms of the divine. Benevolent nature features prominently as a source of such symbols that substitute something tangible in place of the inexpressible. As an example, Eliade writes about the 'Cosmic Tree, whose roots plunged down into Hell, and whose branches reached to Heaven'.[55] The symbol appears in Indian, German and Chinese mythology, and is also the axis of the world, and a ladder for traversing between levels of being and awareness. For Eliade, 'All symbolism of transcendence is paradoxical, impossible to conceive at the profane level.'[56] Such symbolism speaks of an enchantment that exceeds the magical discourses of fantasy narratives, and the quasi-science of science fiction. It is in many respects an ordinary part of nature and the life world.

Disenchantment

According to the early sociologist Max Weber, 'The fate of our times is characterized by rationalization and, above all, by the "disenchantment of the world".'[57] Griffin identifies Weber's term Entzauberung which means 'taking the magic out'.[58] As I have shown, putting the magic back in, re-enchantment, is a theme of some interest among contemporary digital developers and scholars. But the genie is out of the bottle – or has evaporated. Scientifically minded intellectuals will accept that striking two stones together will make fire, but they are unlikely to return to the view that elixirs from the fairies have the same curative powers as aspirin, or that animal sacrifice will increase the harvest. In those days that was just the way the world was. Prior to Shakespeare's time, rituals, incantations, watching crops grow and fire were all just as ordinary or mysterious. People did not relate cause and effect so precisely. Science was just nascent. There was no real magic either. In fact, many have proposed that magic is a product of modernity

> predominantly made to define an antithesis of modernity: a production of illusion and delusion that was thought to recede and disappear as rationalization and secularization spread throughout society.[59]

There are many other explanations (discourses) about how things interrelate without recourse to fields of enchantment. I have attempted to show in this book how the ubiquitous communication of signs provides such a unifying narrative. Much depends on language. Martin Heidegger explained being-in-the-world that seems superficially to endorse enchantment:

> When I go toward the door of the lecture hall, I am already there, and I could not go to it at all if I were not such that I am there. I am never here only, as this encapsulated body; rather, I am there, that is, I already pervade the room, and only thus can I go through it.[60]

He was not claiming a paranormal power to be in more than one place at once, or that he had access to a life force so far undetected by science. He was writing as a phenomenologist. He examined what it is to be in a place, redolent with expectations and fore-projections, and in the process, he examined how to break from a language that presumes we are independent thinking subjects isolated from an object world. Heidegger's is a particularly rich and provocative language game. The quote is from the same essay where he says, 'Language is the house of being.'[61]

The discourse in the empirical literature, even when it focuses on the paranormal, seems to ignore the linguistic turn in philosophy and much of the arts and humanities. It bypasses the insights of phenomenology and semiotics. Trust in invisible biological connections or undetectable force fields is *less* than what language offers. I would also say that trust in empirical observation, measurement and calculation provides the inquiring mind with less than analogy, metaphor and symbolism offer. There is no escaping the language and culture of inquiry. As well as phenomenology, there is poststructuralism to contend with for which magic is not something to be asserted, verified, defended or refuted, but has a political and linguistic import. Derrida talks about magicians as outcasts in *Plato's Pharmacy*.[62] There the discussion is not about whether magic exists, but the way Plato uses language to explain how any claims to authority and centrality inevitably deploy that which is at the margins. So, the magicians were outcasts living on the edge of the city, but were also retained at the central temple area ready to be sacrificed to the gods in the event of an impending siege. Magic comes to represent something that is outcast but already in our midst.

The new age re-enchantment of nature makes for compelling stories and quests. It does however sideline the complex cultural and social factors that govern the character of a place. In the search for magic and mystery it is easy to ignore the prosaic and the obvious. But then most of us can discriminate between fantasy and the everyday. The two seem to coexist.

Mundane places

Magic is also a surrogate for imagination. Some places excite the imagination. Who would deny the value of a nature setting that suggests wood nymphs, a mountain that could speak and a lake likely to yield up Excalibur? Isn't a world bereft of such imaginative entailments and magic boring? It is worth concluding this discussion of numinous nature with a slight digression that brings us back to nature. We need to plumb the depths of banality to appreciate what it is we mean by nature.

Consider boredom. Is it a thing of the past? Since I acquired a smartphone there has been no such thing as 'down time'. Ten minutes waiting for a bus used to seem like an hour. Now it is barely enough time to check an email. The *Guardian* featured an article about what we gain by recapturing empty time, when nothing happens and we just wait. Boredom can be good for you. Apparently, 'Aimlessness, rest and even boredom can boost creativity.'[63] But Martin Heidegger proposed that there are other ways to deal with boredom, and at least three ways to be bored.

First, you can be bored by something, for example, waiting for a train in an isolated station and with nothing to do. In this case time draws attention to itself and is slow and conspicuous. The person waiting, checking timetables, staring at their watch and pacing up and down, is removed from engagement, unable to be immersed and in limbo. The place is not boring, but in Heidegger's terms the drag of time denies the train station the opportunity to offer the would-be passenger anything of interest. Heidegger says that in a sense the station presents itself at the wrong time. One commentator, Espen Hammer, summarizes this condition: 'Time itself does not bore us; it is the essential being held in limbo in coming to be left empty which constitutes the first form of boredom.'[64] Second, you can be bored with something. You are engaged in some event, but perhaps afterwards realize that you were in fact bored, though not with anything in particular. The company, the entertainment and the food kept your interest. For Hammer, summarizing Heidegger, this is 'the empty "lived-experience" of entertainment and distraction'.[65] Perhaps this is the condition of watching videos, flicking through your playlist, reviewing emails, deleting spam and checking 'likes' on Facebook just to fill in time.

But Heidegger is really interested in the third kind of boredom that is with nothing in particular. He calls this *profound boredom*. This is a condition of complete indifference. The world falls dead. We expect nothing of it. We might just say, 'It's boring,' and even, 'I'm bored,' though even that does not quite capture the condition. Heidegger says enigmatically that this kind of boredom 'is the entrancement of the temporal horizon, an entrancement which lets the moment of vision belonging to temporality vanish',[66] a position that he elaborates over many pages.[67] Whether Heidegger's elaborate argumentation is to everyone's taste, there is an interesting tactic in play here. To elevate boredom in this way is to consider something we normally consider as a deviation from the ideal (i.e. boredom as a deviation from the ideal human condition of being really interested in the world around us) and to reverse the priority. If someone complains to us that they are bored, that the world has lost enchantment for them, then we can rejoinder that they are not bored enough – you really need to plumb the depths of profound boredom. Sink into the depths of nature's banality. Strip it of magic and imagination, at least for a time, and harvest the riches of its semiotic relationships.

I started this chapter with the case for magic in nature settings. Nature appears as a network of invisible organic connections that defy overt detection, but that exert influence nonetheless. Contemporary enthusiasts of fantasy and science fiction exercise a primitive urge in all of us to identify such connections in nature. I looked at quasi-scientific concepts such as morphic fields, electricity as life force, and how video games and other digital media play with and reinforce such narratives. We also play in the so-called magic circle of the game. I alluded to spatial paradox as displayed in games, art and architecture, that demonstrate further our propensity to construct alternate spatial realities. I touched on symbolism as a component of semiotic play. The numinous qualities of nature spaces seem to depend on symbols. Symbols bind communities together and to the landscapes they inhabit. Then I reviewed the concept of disenchantment, as if we have lost something in our reading of nature, leading to a discussion of mundanity and its potential. In the next chapter I review how some of us digital operatives disconnect from each other, society and the rest of network nature.

CHAPTER 11
NATURE UNPLUGGED

Self-reliance is one of the signs of mature adulthood. By most accounts self-reliance takes time to develop. It is in the company of health, well-being, morality, aesthetic quality and other virtues and 'goods' that people also associate with nature. As attested by many nature writers, nature comes to stand in for self-reliance, and acts as the site in which our independence is most clearly exercised.[1] The American transcendentalist Ralph Waldo Emerson (1803–82) took nature as the model for the indomitable human spirit.[2] Anticipating Martin Heidegger[3] on the poetics of letting-be Emerson observed that the flower growing beneath his window does not need to justify itself, to compare itself with other flowers, or flowers from the past.[4] Outdoor pursuits that bring people into contact with nature also fall into the orbit of 'rational' outdoor recreation as promoted by nineteenth-century reformers intent on encouraging the working classes to do something healthy with their spare time. Outdoor sports, walking in the countryside and more adventurous pursuits are good for you and promote independence.[5] You learn to rely less on the comforts of city living and its attendant support structures – including minders, helpers, guides, counsellors, authority figures and the state.

Advocates of self-reliance indicate that technologies can confound this journey to independence. In the face of our increasing dependence on digital technologies it gets harder to maintain the illusion, let alone the reality, of self-reliance. So, there is a conflict. Those who believe in the primacy of the individual – self-sufficient and independent – become aware of how much we depend on transportation systems, networks, smartphones and other paraphernalia of the modern world. In an article on the theme of biology and architecture, Andrew Ballantyne reminds us that we inhabitants of the city 'are now routinely without fundamental survival-skills, and would be lost without access to clothes and vehicles that we do not know how to make, as well as shelter and electricity that we know well enough how to use but only once they are in place'.[6] Among all the technologies on which we rely, our phones and other personal devices demonstrate a condition counter to self-reliance – technological dependence. No wonder people have mixed feelings as they depend on their devices.

From my own work with digital technologies, I observe that there are at least two ways that digital devices upset people's ideal of self-reliance. The first is via the functions provided by such technologies – information, communication, navigation and countless other app functionalities, not to mention all those commercial, industrial and big-data systems and infrastructures on which we all depend. We cannot do without them, or at least we persuade ourselves that they are indispensable, to the extent that many commentators use the language of *addiction* to describe the human–technology relationship.[7] The second affront to self-reliance is the individual's increasing dependence on technical support from other human beings. It is not just that you must learn how to select, purchase and use these devices, but there are upgrades to download and install, new access protocols to negotiate, new peripherals and features to purchase. The technical support dependency applies not only within organizations, but also for the individual consumer. The situation is particularly acute as people get older and confront changes that overturn a lifetime of habits, as testified in the growing field of gerontechnology.[8] Calling on tech-support challenges our self-regard and our sense of self-reliance. In what follows I will tease out some of the characteristics of self-reliance in the digital age, starting with the proposition that we ought to depend less on digital systems.

Being post-digital

The idea of the post-digital captures an ethos in which people consciously go off grid to regain access to the natural and the authentic, and become self-reliant. The term *post-digital* has its origins in the reflections of Nicholas Negroponte, author of *Being Digital*. In 1995, he asserted that the online digital world will surpass the constraints of geography, that 'the digital planet will look and feel like the head of a pin'.[9] In 1998 he stated in *Wired* magazine that 'the digital revolution is over'.[10] This is a kind of digital severance. We no longer need to speak, advocate for, or on behalf of, the digital. Digital technology is so woven into the fabric of everyday life that it no longer needs a special label, nor social commentary, nor propagandists (like Negroponte). It has gone the way of the plastics business – once regarded as revolutionary, but now taken for granted and unremarkable.[11]

The term 'post-digital' has its genesis in sonic arts and music making around the year 2000. Adopting Negroponte's message, sound artist Kim Cascone justified his mixed media approach to music making.[12] This was *post-digital* culture. Composer-performers could bypass technical theory about digital signal processing (DSP): 'Sometimes, not knowing the theoretical operation of a tool can result in more interesting results by "thinking outside the box"'.[13] They would embrace the tools, and their peculiarities, and would fixate less on expert digital knowledge. In the world of sound, the adherents of post-digital cultures relaxed the idea that digital technologies need to access precise, faithful and true representation and expression. They did not have to mimic nature. Just as people enjoy old media's quirks and imperfections, so too an artist can work with digital glitches.[14] Knotted wood grain, squeaky guitar frets, peeling paint, the patina of age and the textures of natural materials have their digital equivalents. Such insights inspired the way of looking at technology and nature as *post-digital*.

Communities and cultures imbued with a post-digital ethos place digital media among other media. You need the right tools for the job. Do not go for the hi-tech option just because it is available – and certainly not to impress your audience. Post-digital culture is not against digital technologies, though it continues various anti-industrial strands from the arts and crafts movement and romanticism. Now magazine journalists write about post-digital cultures, advertisers take note and adopt the terminology and there is an online peer-reviewed journal called *Postdigital Research*.[15]

There are critics of the post-digital and, as a self-reflective study, it generates its own internal critique. In his article 'What is post-digital?'[16] Florian Cramer proposes that post-digital cultures exhibit 'either a contemporary disenchantment with digital information systems and media gadgets, or a period in which our fascination with these systems and gadgets has become historical'.[17] There always have been sceptics. Now we can locate the former fascination with digital technologies in history. For Cramer, the post-digital devotee seeks out an ever-elusive 'authentic' experience. A 2014 *Guardian* article about post-digital cultures references the performance artist Marina Abramović. In a YouTube interview, she shows us how to savour a glass of water.[18] The post-digital indicates a trend to go live, to attend concerts in the flesh and to be in the moment. Real life is not online. But neither is the post-digital obsessed with digital skill and knowledge. Why speak any longer about digital architecture, digital medicine, digital ethnography, digital engineering, digital writing, digital aviation, digital art or digital nature? The digital is everywhere. According to Cramer, '"Post-digital" thus refers to a state in which the disruption brought upon [*sic*] by digital information technology has already occurred.'[19]

According to Cramer, heirs to the post-digital age also reject 'the kind of techno-positivist innovation narratives'[20] found in *Wired* magazine, the 'singularity' movement (to be discussed below), and Silicon Valley. They also reject 'the Quantified Self movement, and sensor-controlled "Smart Cities"'.[21] I would add 'parametricism' and 'biophilic design' (Chapter 5) to this list. The world is not big data. The post-digital practitioner chooses media and tools suited to the task at hand, unselfconsciously. Cramer starts his article with the image of the hipster with a typewriter in the park offering to type up personalized stories for passers-by. The hapless writer provides a nice post-digital motif as it is an art performance, delivered with irony, and indicates a technology choice more suited to sitting on a park bench surrounded by nature with no printer or power supply to hand: 'A post-digital choice: using the technology most suitable to the job, rather than automatically "defaulting" to the latest "new media" device.'[22]

The post-digital enthusiast prefers DIY to factory-produced objects. Champions of the post-digital dispense with distinctions such as new media versus old media, or digital versus analogue, but they are aware of the distinctions between DIY versus corporate, independent versus global multinational and militarist. Perhaps there was a time when it made sense to speak of the internet as a radically permissive medium, a wild frontier and a truly democratic medium. But those days are over. There are resonances here with biohacking culture I described in Chapter 4. The post-digitalists of course use the internet and whatever media are to hand to exercise their commerce in small-scale, independent, DIY, semi-crafted practices and works. But they know that big corporations do this too.

Cramer identifies a 'semiotic shift to the indexical'[23] and away from symbols. The indexical relationship is a direct connection between the thing and its sign. I take this to mean an emphasis on collections and their materiality, without resorting to mediation through an overarching ordering system or theory. This reminds me of the flea market,

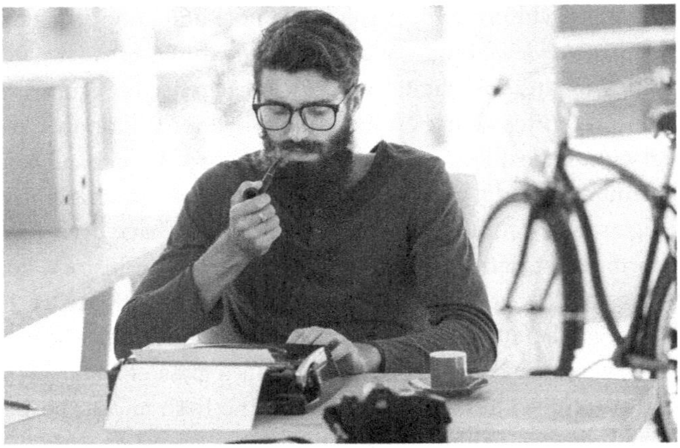

Figure 11.1 Hipster with typewriter.
Source: Shutterstock.

cabinet of curiosities and back rooms of a museum. Just show me what you have got – never mind the smart display, the meanings and complex narratives. There is also a link here with Chris Anderson's (editor of *Wired* magazine) proclamation about the 'end of theory' (Chapter 3), and picked up by David Berry and Michael Dieter.[24] Never mind the theories, let the data speak for itself. This is an exception to the proposition above rejecting techno-positivism. The post-digital is not without contradictions.

The post-digital also embraces the aesthetic. There is a subtle argument here that the digital cuts out the aesthetic. For Cramer, 'Our senses can only perceive information in the form of non-discrete signals such as sound or light waves.'[25] So, 'anything aesthetic (in the literal sense of aisthesis, perception) is, by strict technical definition, analog.'[26] Post-digital cultures reject the proposition that the world is made up of digital information (bits and bytes), and they thereby reclaim the aesthetic. Post-digital aesthetes want to reclaim agency (control) over their lives. The quest for agency of course extends to others, and to supporting others in the same quest. We need to be in control rather than submit to the pressures of advertising, big government, corporations, militarism, colonial rule and class categories. Post-digitalists marshal the internet and mobile communications in the exercise of agency, while recognizing that others also use such tools to control us.

Post-digital cultures embrace agency, self-determination and self-reliance, but from the political left, as it were. Cramer says that the post-digital suffers from the same delusion as other ideologies, that we can be in control of our own destinies (i.e. we are autonomous agents). After all, the 'post' prefix does not mean the abolition of the old power structures, but their transformation: 'mutation into new power structures.'[27] Cramer defines the post-digital as a 'term that sucks but is useful.'[28] 'Post-digital' is a useful term for me in this book while considering the relationship between the digital and the natural. In summary, the post-digital is a return to a certain kind of nature, constructed unselfconsciously, and with technology still in the frame. Deploying digital technology need not make any practical difference to how we regard the world and what we value.

The accessorized self

The iconic post-digital image of the hipster with a typewriter is now replaced by the regular post-digital millennial with a smartphone.[29] To develop 'self-reliance' is to shift reliance structures from the familial world of parents and guardians to teams of peers and contexts of mutual support, and to assume a demeanour that complains less, takes responsibility and gets on with the job at hand.[30] The term 'self-reliance' comes with qualification. Individuals at some stage need family, support communities and institutions to survive and thrive, and everyone deploys tools and technologies. In fact, myths of self-reliance frequently place the heroically independent individual in the company of at least one iconic accessory. As well as the hipster with typewriter, there is the lone troubadour with lute, the samurai with sword, the cowboy with a six shooter, shepherd with her crook and Artemis with her bow. To carry a smartphone is not alien to the independent spirit, though its utility as multifunctional communicative 'Swiss army pocket knife' amplifies the capabilities of the accessorized self-reliant in ways that are new and of consequence.

Self-reliance has become a catchword of both the political left and the right. The hipster mindset is liberal, left-leaning and informed arguably by art practices, a sense of history, concern for the environment and intellectualism. For the political extreme right, self-reliance means independence from city-based intellectualism, and is suspicious of the state. It advocates living off grid, but not without tech-accessories. Down-to-earth self-reliance reveals itself in the folksy and family-friendly advice from *Self-Reliance* magazine for the so-called 'prepper', the enthusiast who wants to be able to survive if the social and technological machinery breaks down. According to one prepper correspondent:

> One of the best forms of personal protection is living where you probably won't need it! We live three hours north of the largest urban area, in a remote, off-the-beaten-path location, separated from the road by 1.3 miles of dirt trail. We have large dogs, weapons, and the experience to use them. Do you?[31]

The 'prepper's' accessory kit includes weaponry:

> Be sure you have adequate ammunition for each of your guns on hand, stored in airtight, waterproof containers. Damp ammunition or the wrong ammunition is dangerous – it's worse than no ammunition![32]

For the prepper, if you cannot live in splendid and remote isolation in the countryside then you still need to prepare in case you must hunker down or beat others in the rush to escape after a tsunami, earthquake, terrorist attack or breakdown. In any event, *Self-Reliance* magazine advises:

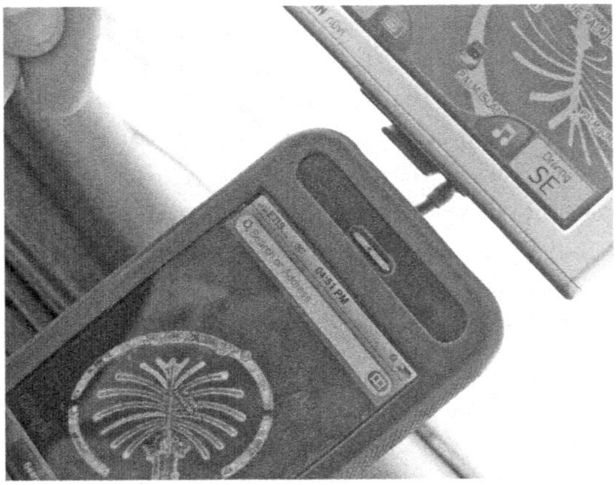

Figure 11.2 Self-reliant's smartphone and satnav for navigation, Palm Islands, UAE. *Source*: author.

> Have a good pocket knife in your pocket at all times (everyone in our family carries one) and a lighter, as well. And have a couple of quality kitchen knives and a pack of a dozen lighters in your kitchen gear, as well. Add a sharpening stone, a little dry tinder (just in case), and your kit is in good shape![33]

Living in the countryside, the woods or on a farm is ideal for exercising this kind of anti-state self-reliance.

Emerson, the nineteenth-century transcendentalist, held the state and government in contempt, a sentiment on which current right-leaning populist politicians draw.[34] Whatever his political persuasion, he seems to have been taken up by the far right in the United States.[35] There is no evidence that Emerson took his self-reliance into the countryside. He thought it wonderful to enjoy the spectacle of nature in quiet armchair contemplation, but he thought the skill is to exercise the same sense of independence and solitude while in the heart of the city and its commerce: 'The great man is he who in the midst of the crowd keeps with perfect sweetness the independence of solitude.'[36] It was his friend Henry Rouseau who left the city to exercise self-reliance in a rustic hut by the lake at Walden,[37] though there are ample critics of the compromises and failures of that project.[38] Emerson believed that we should be independent individuals, nonconforming and self-reliant. He wrote: 'Whoso would be a man must be a nonconformist.'[39]

Narratives of self-reliance of all political colours inform much of the discourse today about the human's relationship with nature. People's apparent addiction to the internet, social media and video gaming is also supported by such narratives. The search for new kinds of digital currency that break with the need for banks and other financial institutions also speaks of the goal of self-sufficiency. For cultural critic David Golumbia, digital currencies 'emerge from the profoundly ideological and overtly conspiratorial anti-Central Bank rhetoric propagated by the extremist right in the U.S.'.[40] As for cash-only-economies, digital money serves anti-establishment preppers suspicious of the 'deep state', and other putative adversaries of the alt-right.

I also suspect that contemporary anxieties about artificial intelligence, super surveillance, autonomous self-replicating machines, machine domination and the so-called singularity cluster around the blatant impossibility of self-reliance. Contrary to the myth of self-reliance, human beings have always been in societies, organized themselves and relied on the technologies they invent. It need hardly be said that steps towards 'prepper' self-reliance are dependent on, if not parasitic upon, infrastructures, institutions, systems, industrial production, networks, economic realities and society in general.

Self-reliance does not scale up as a social model. It does however signal an anxiety: what happens when our lives are so inextricably bound up with technology that technology constitutes a life-support system and we cannot survive without it? Then what happens when it all comes to a stop? How can I cope when left to my own resources, but without the machine? Apart from the apocalyptic distress of society in ruins, there is the challenge of confronting nature in the raw – when zoo animals go feral,[41] the triffids take over,[42] apes keep humans as slaves,[43] and other post-apocalyptic science-fiction scenarios.

Simulation

To depend completely and utterly on technology gives rise to a fascinating series of thought experiments. If human life is accessorized by machines, then what of the reverse, where all that we regard as human is somehow supplemental to the operations of the machine? This is the basis of one of Karl Marx's critiques of the industrial age and the capitalism that supports it.[44] Human labour becomes the disposable adjunct to the operations of mechanization and industrial production. This scenario receives more literal and fantastical treatment in films such as *The Matrix,* where human bodies are stored in silos as a power supply for intelligent machines, while the human minds attached to those bodies are preoccupied with a massive, shared simulated existence in something that seems like the real world. The separation of mind from body features in popular thought experiments, as in the philosopher Daniel Dennet's proposition about a brain existing in a vat of nutrients independent of the human body, with complex wiring and radio hook-up to its fully functional body some distance away from it.[45] Cognitive philosophers are intrigued by 'the brain in the vat' as a thought experiment.[46] The brain upload idea reached its apogee with Hans Moravec's concept of an ultimate mind meld as all of nature gets absorbed into a cosmic mind bubble.[47] Bodies are not necessary for the simulation idea to have traction. If the universe is all data and code, then human minds can exist independently of bodies in the circuitry of some future electronic mega-brain.

There are philosophical consequences to the big-data scenario outlined in Chapter 3, and human dependence on the machine. If data and programs can represent the entire universe, then it indeed seems plausible that software, hardware and the appropriate interfaces develop to the point where a simulation is indistinguishable from the lived world, and may even exceed it in complexity. The simulation idea also features in models of how we interact with the world. For example, ideas by philosopher and cognitive scientist Donald Hoffman have achieved some prominence. He proposed that what we perceive as the world about us is already 'an illusion' as it is geared towards fitness in the game of survival rather than some reality beyond our perception:

> According to evolution by natural selection, an organism that sees reality as it is will never be more fit than an organism of equal complexity that sees none of reality but is just tuned to fitness. Never.[48]

In other words, even if our view of the world is delusional, if it helps us survive and thrive, we are fine. Having an 'accurate' perception of the world hardly matters, and may provide too many distractions anyway.[49] I mention this Darwinian argument as it is couched in terms of code theory and mathematics, in other words in terms of arguments that have traction in the world of big data.

Once you start to think in such data-oriented ways, so the narrative goes, it is plausible at least to conjecture that we are living in a simulation even now, as in *The Matrix*. As a thought experiment, philosopher Nick Bostrom proposed that we might now all be living in such an artificial world contained in a digital simulator.[50] He uses that idea in a thought experiment to test ideas about ethics, knowledge, memory, our relationships with one another and with nature and about other questions that engage philosophers. It is worth pursuing the simulation scenario for what it says about the human response to nature. According to the scenario, thousands of years from now, our descendants, equipped with unimaginably superior technology and knowledge, and millennia of history to look back on will decide to create the ultimate re-enactment of history on planet Earth. It is like a costumed re-enactment of the Battle of Hastings, but on a computer, as in *The Sims* or *Civilization*, and on a vast scale and with intricate detail, requiring a computer the size of a small planet. Apparently we inhabit this simulation and exist millennia into the future, even though we think we are in the twenty-first century.

According to the scenario, superior, future, transcendent humans (post-humans) might have several motivations for creating this simulated universe: to aid historical inquiry, to use as a digital time capsule or a game or a reality TV show entertainment, to learn from the mistakes and successes of the past, to respect and show veneration for ancestors or for the sake of nostalgia. In his paper Bostrom assumes that such a computer would need to contain all the laws of physics, descriptions of natural systems, materials, forms, shapes characteristics and of course the cognitive and life processes of all living things, as well as the nonliving. He modifies this Herculean project by suggesting that some data, such as information about distant planets, the centre of the Earth and sub-microscopic entities could be at much lower resolution or generated by the simulation on 'an as-needed basis'.[51]

I would have thought that simulating the cognitive capacity to dream and recall would be sufficient to keep the simulation going, as in the film *Total Recall* (1990).[52] Bostrom does not explore this possibility. That might be too difficult even for our god-like post-human descendants to observe as they seek to understand what made their ancestors (i.e. us) tick. In any case the hallucination of reality must be shared by all the sims in the simulation. Unlike independent dreamers, the actions of one influence the actions of others. If you think such a simulation is vaguely plausible, you may as well entertain the possibility that our computer-savvy descendants are themselves in a simulation invented by their descendants. So, there are simulations within simulations, including false starts, defective simulations (like those early Egyptian proto-pyramids that collapsed) and alternative realities.

Such fantasies seem to play off at least five deeply ingrained contemporary cultural threads, giving them a high-tech twist. First is the data model of nature to which I have alluded a few times, and the neoplatonic idea that there are layers to existence, with ultimate reality lying somewhere beyond our current condition in the world.[53] Second, the simulation hypothesis draws on the related interpretations of George Berkeley's (1685–1753) philosophical idealism that matter and nature depend for their existence on human thought and consciousness.[54] There are no material substances. Third, the ancestor-simulation hypothesis also harbours the empirical view of history: that we crave the facts, the indisputable original truth about what happened years ago.[55] If we can recreate history then, at last, we will know what happened and what it was like to be there. Fourth, anomalies and inconsistencies in science, observation and life generally find some explanation in the idea of imperfections in the simulation. Such anomalies extend from the behaviour of particles at the quantum level to the existence of unaccountable coincidences, action at a distance, ghosts, magic, out-of-body experiences, the numinous spaces described in Chapter 10, and even injustices. At last the universe makes sense. Fifth, there is infantile egocentrism that most of us grow out of. This is the child's suspicion that he is the only one alive or conscious while everyone around him is an automaton or actor in a play with the child as the focus. In the latter narrative, the child even wonders if he will ever catch out these actors, or discover a bevy of operatives manipulating everything behind the scenery as in *The Truman Show* (1998).

Affirming, playing with and debunking the simulation hypothesis have become something of an academic industry.[56] Bostrum affirms the widely held data-code position that 'it is not an essential property of consciousness that it is implemented on carbon-based biological neural networks inside a cranium: silicon-based processors inside a computer could in principle do the trick just as well.'[57] Leaving aside the question of consciousness, if you adopt the phenomenological position of 'embodied cognition'[58] then bodies, contexts and worlds are crucial in the way we humans think, react, interact and are in the world. To deny the physicality of the world makes good science fiction, but is impractical and founded on a very restricted premise of what cognition is.

For me the simulation hypothesis provides a service in that it reduces to an absurdity the data-ized view of nature and of biological existence. The second outcome is that it shifts our anxieties to another forum: What happens if someone pulls the plug on this simulation? Unlike in *The Matrix*, in this ancestor-simulation there are no actual bodies to jump back into. Though if future post-humans are clever enough to produce the simulations in the first place then perhaps they are smart enough to download our minds into whatever their equivalent is of bodies. But then they might not bother. It is all just bits and bytes after all.

Other science-fiction scenarios also deal with the anxiety of nature unplugged. The classic science-fiction short story by E. M. Forster, *The Machine Stops* (1909) summarizes the common anxiety about what it is to be absorbed into technology, and delivers the narrative through a dialogue between a mother and her son (Kuno) in a future where everything is controlled by 'the machine'.[59] People lounge in armchairs in their hexagonal hive-like pods listening to lectures and having 'ideas'. Everything is synthetic. Air travellers close the blinds on their airships lest they catch a glimpse of the Alps and sunlight. Kuno's mother says, 'I dislike seeing the horrible brown earth, and the sea, and

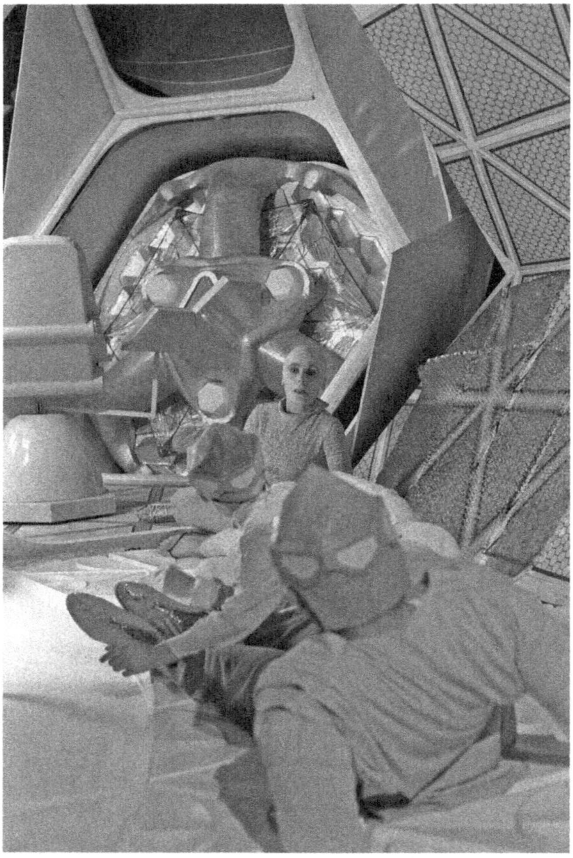

Figure 11.3 Final scene from the 1966 BBC television adaptation of E. M. Forster's *The Machine Stops*, episode 1 of series 2 in the series *Tales of the Unexpected*. © BBC 1966.

the stars when it is dark. I get no ideas in an air-ship. ... What kind of ideas can the air give you?' The story ends with an airship crashing into the honeycomb that was the city. The last thing the mother and son see is the 'untainted sky'.

The story serves as a warning against putative loss of individuality and self-reliance, pushing to extremes their opposite: mindless acquiescence to the power of something else – the machine, the collective, dictatorship and state capitalism. In Forster's story the villain is the all-pervasive electro-pneumatic machine and people's acquiescence and conformity, but he could just as well be referring to the machine of the state. Contrary to Emerson's idea that nature is the model for the indomitable human spirit, the conceit among the inhabitants of Forster's dystopia is that nature has been tamed and rendered obsolete. It has nothing to teach us.

If the machine were to fail, could we rely on just ourselves? That is a dilemma for someone who takes for granted the innate autonomy and independence of the human being, and someone who can only imagine the alternative to self-reliance as a hive-mind existence in a machine-led dystopian world. That makes compelling fiction, but the world as lived seems to be otherwise, and much more complicated.

Metonymy and nature

Permit me to outline some semiotic and linguistic theory to situate people's advocacy for self-reliance. Simulations bring to light the role of metaphor in human understanding. To see the mind as resident in brains and as transferable between bodies and machines is to adopt a particular view of mind, and also to give prominence to a particular metaphor: the mind as substance or fluid that can be poured from one container to another. We do think metaphorically after all.[60] Neither is nature immune from metaphor, and throughout this book I have drawn substantially on the network metaphor, and in turn granted prominence to language.

To return to the scenario of the digital self-reliant, let us cast the human–machine relationship in terms of language. A metaphor is simply a statement in which we see one thing as another: a man as a wolf, a tree as a giant, nature as a network. Metonymy is a subspecies of metaphor. A metonym is a smaller part of something larger used in language to represent the whole, as when 'crown' stands for the monarch, 'pen' stands for the writer and 'sword' stands for the warrior. It helps to consider any technological accessory to the human as metonymic for the individual, as if the accessory defines the individual. By extension, the accessory also stands for something larger, the individual's personality and values, the 'equipmental whole' within which the human being is immersed,[61] the techno-social sphere on which the individual depends – and that made the individual who she is. Whatever a good penknife means to a self-reliant 'prepper' the accessory also speaks of its manufacture and of the cultural practices in which it is situated (whittling, repairing, hunting and self-defence). If such connectedness applies to objects with limited functional entailments, then how much more can be said for the self-reliant's smartphone. That particular accessory renders obvious, and in a literal way, just how connected we are, especially when it breaks down or goes missing. The breakdown, loss or inadequacy of the self-reliant's iconic accessory is not just a loss of function, but a loss of identity and a breach in the web of connections. As Heidegger suggested in the case of the carpenter with a defective hammer,[62] breakdown brings the object into awareness as a thing, and brings to light the network of dependencies, the reality of the interdependent social and technological whole and the importance of resilience in the face of potential loss.

Clothes and other wearables are also accessories. In her investigation into women, shoes, purses and fetishes, cultural theorist Elizabeth Halsted concludes,

> These particular items of clothing have rich symbolic meaning, metaphorical powers, and serve important defensive functions which ward off painful or fragmented thoughts, contain affects and sensations, or connect their wearers to powerful or positive self-states.[63]

I would add that whatever Freudian or Marxist reading one wishes to place on clothing applies also to other symbolic and functional accessories: guns, staffs, utility belts, capes, wands, sonic screwdrivers, whips, drawing pens, typewriters (which are the accessories of Wyatt Earp, Gandalf, Batman, Superman, Harry Potter, Dr Who, Wonder Woman, architects and hipsters) and smartphones. Accessories are props for whatever role we wish to assume in life, and they reinforce our identity. The accessorized human suggests independence and self-reliance, but on reflection we see that accessories are simply stand-ins for an interconnected whole. The prepper with a penknife depends on the overall socio-technical system through which the accessory is manufactured, distributed and acquired, as well as the processes by which practices are passed from parents to children; cultural memes are therefore formed and supported by economic, legislative and educational systems. Humans depend on social and cultural systems as they depend on their accessories. We cannot do without each other, technologies, world and nature.

Denatured

Deep ecologists and lovers of nature may well subscribe to the view that nature is one, a whole, a unity. As I have mentioned a few times, that is a view supported in various guises through existential monism, a movement to which many of the romantics subscribed, along with pragmatists such as C. S. Peirce. As I have attempted to show, in the digital age we tend to interpret such unitary understandings via network models. Biology informs us that the elements of nature are highly interconnected. We extricate the components of nature from one another as if they are independent, objectifying them for the purposes of explanation, analysis, diagnosis, prediction and control, and other aspects of instrumental thinking. From the point of view of the existential monist we humans are at heart so interconnected, so networked to one another, to earth systems and nature that the autonomous, independent human individual is but a convenient fiction of the moment, and of language.

That is a reasonable approximation of the relationship between the individual and the totality of which they are a part, and language all but fails us on this point. Another tactic is to play with the 'movement of negation'. The environmental philosopher Tetsuro Watsuji (1889–1960) expresses our interdependency thus:

> An individual who does not imply the meaning of negation, that is, an essentially self-sufficient individual, is nothing but an imaginary construction.[64]

The idea of the individual predicated on self-sufficiency is a fantasy. One could also say in negative terms that we are neither autonomous individuals nor indistinguishable within a totality: 'These two negations constitute the dual character of a human being.'[65]

A simpler variant of this negative discourse is to define networks by their disconnects. Digital communications technologies provide a convenient model of what networks are like, and how networks are 'susceptible to disruptions that spread through the underlying network structures, sometimes turning localized breakdowns into cascading failures or financial crises', according to a review by David Easley and Jon Kleinberg.[66] Espionage, subterfuge and sabotage infiltrate and disable network structures of communication, transportation and commerce. Certain online activity persuades individuals to disconnect as well. A cursory glance through the Twitter streams of high-profile micro-bloggers reveals online commentaries that intrude, criticize, threaten, defame and abuse.[67] It is no wonder that some among the social media community want to switch off and disengage. If we think of nature as a system of connections, then networked nature also inherits this condition of breach and disconnect.

In this chapter I examined self-reliance and the role of digital technology in marking out the autonomous, independent individual. The ambition and anxiety of independence from each other and nature extends across the political spectrum. Self-reliance draws from ambivalent feelings about how much we depend on machines – the anxiety that we depend so much on technology to support life; but one day it will all fail. I think that fear of infrastructure failure eclipses anxieties that the world will be overtaken by artificial intelligence, super surveillance, autonomous self-replicating machines, machine domination, simulations and the singularity. In this chapter I turned to accessorized self-reliants and the role of metaphor and metonymy, which brings us back to language, semiotics and the communicative structures of nature.

C. S. Peirce was operating within and responding to a climate informed by a popular romantic naturalism and the desire to escape the machine and get back to the woods, an impossibility inevitably and thankfully averted and unrealized.

CODA

In the Introduction, I suggested that the natural–digital dichotomy is not an antagonism between equals or opposites, or between two sides of the same coin. Some oppositions do lie at either end of a spectrum, with gradations between, as is the case with light versus dark. That is an opposition with an inverse relationship. The more light you add to a room, the less dark it is. The louder things get, the less quiet; the hotter, the less cold. But *the natural versus the artificial* is not that kind of opposition. More of one does not imply less of the other. To assume such a relationship is to succumb to the *displacement metaphor*, a subspecies of what George Lakoff and Mark Johnson call 'the containment metaphor'.[1] That metaphor assumes there is a fixed space in which to give play to our concepts. The more you fill this space with one thing, the less space there is for another. It also implies balance. Too much tonic and there's not enough space for the gin. The measure does not have to be 50:50, but more of one implies less of the other because the volume of the container (the glass) has a limit. People apply the displacement metaphor to other opposites: if you fill your life with sadness, there is less space for happiness; more love, less hate. These are dubious applications of the displacement metaphor.[2]

Nor is *nature versus artifice* the kind of opposition that lends itself to the metaphor of displacement. The opposition more closely resembles the opposition between hats and shoes. There is something spatially oppositional about *hats versus shoes*, but every time you buy a hat you do not need to throw away a pair of shoes. *Nature versus artifice* is also like the opposition between structure and ornament. Having more ornamentation on a building does not mean there's less structure, though it may be more hidden from view. In fact, adding more ornamentation can mean increasing the amount of structure to hold it in place. As any architect knows, the two are not so neatly defined anyway. The nature versus artifice opposition is much lumpier than the displacement metaphor implies. In any case, nature is not a quantitative thing of which you can have more or less. Nor is it even a thing, but a catch-all convenience with fluid boundaries (to mix in another metaphor).

The threats to environment mentioned in this book are real, complicated and 'wicked',[3] but a raft of threats does not always imply increase from an aggressor. Nature is not driven out by more technology, but by bad policies and practices. More of one thing does not need to drive the other to extinction, notwithstanding energy-consuming and heat-generating data centres and infrastructures. As a prime example of artifice, I have focused on digital technologies. There is ample evidence that digital devices and systems help define and redefine what we mean by the natural, and with no necessary detriment to its object. Digital technologies also constitute tools and systems for accessing what we think of as nature. The nature–artifice relationship is complicated. My caution throughout is to avoid subjecting everything to data and code. Nature is not data. Theories of signs help deflect us from the inevitability of such an instrumental view.

Predictions

Consider the natural formation of a flock of birds circling overhead. It might herald changes in the weather, a disturbance in the landscape or an approaching army. A skilled interpreter of nature might plausibly identify causal connections between signs, events and their effects on people, other animals, places and things. In contemporary terms, changes in bird populations, migration patterns and flocking behaviour are among the many signs of global climate change. The Latin verb *inaugurāre* is 'to take omens from the flight of birds', according to the *OED*. To say, 'things do not auger well for the future', is to read the flight of birds as predicting misfortune. In Alfred Hitchcock's film *The Birds* (1963) the birds both predict and deliver disaster. Something similar happens in the case of Stephen Spielberg's *War of the Worlds* (2005), as gigantic alien war machines disturb and attract flocks of birds.

Artists and scholars have recruited birds, among other animals, as arbiters of truthfulness. The Roman natural philosopher Pliny the Elder (23–79 AD) describes how the theatrical set paintings of one artist 'were greatly admired, as crows were seen trying to alight on the roof-tiles, deceived by the realism of the painting'.[4] He provides a similar account of birds deceived by an artist's representation of grapes.[5] The art theorist W. J. T. Mitchell deploys these references in an interesting account of the silent and inscrutable ability of animals to adjudicate on matters of truth.[6] It is not that non-human animals are deceived easily, but by their exceptional behaviour they can signal that an artist is delivering an honest representation. Rarely can an artist achieve this level of truthfulness.

Truthfulness is also under challenge on the world stage. The verb *to inaugurate* originates from *inaugurāre*. The start of 2017 saw the *inauguration* of a US administration that seemed to display mocking disregard for truth, and no regard for nature and the environment. During the inauguration ceremony, the Whitehouse website switched over to the new administration's website and 'An America First Energy Plan' that showed a commitment 'to eliminating harmful and unnecessary policies such as the Climate Action Plan and the Waters of the U.S. rule'. This and other disastrous moves have mobilized a groundswell of resistance out of environmental melancholy, and inaugurated a renewed resolve to address the dynamic relationship between nature and technology.

NOTES

Preface

1 Bill McKibben, *The End of Nature* (New York, NY: Random House, 1989).

2 Bill McKibben, 'The End of Nature', in *The Norton Book of Nature Writing: College Edition*, ed R. Finch and J. Elder (New York: W. W. Norton, 2002), 1120–30, 1126.

3 Ibid.

4 Ibid.

5 Ibid., 1127.

6 Ibid.

Introduction

1 See, for example, the antagonism between the natural and the artificial as outlined by Judith Butler, *Gender Trouble: Feminism and the Subversion of Identity* (New York: Routledge, 1990); Donna J. Haraway, 'A Cyborg Manifesto: Science, Technology, and Socialist Feminism in the Late Twentieth Century', in *Simians, Cyborgs, and Women: The Reinvention of Nature* (London: FAb), 149–81.

2 Larry D. Rosen, *iDisorder: Understanding Our Obsession with Technology and Overcoming its Hold on Us* (New York: Palgrave MacMillan, 2012).

3 Eva M. Selhub and Alan C. Logan, *Your Brain on Nature: The Science of Nature's Influence on Your Health, Happiness and Vitality* (Hoboken: John Wiley, 2012), 138.

4 Benjamin H. Bratton, *The Stack: On Software and Sovereignty* (Cambridge, MA: MIT Press, 2015), 35.

5 Ralph Waldo Emerson, *Essays by Ralph Waldo Emerson* (New York: Dossier Press, 2016), 76.

6 Henry David Thoreau, *Walden: Or, Life in the Woods* (Princeton, NJ: Princeton Paperbacks, 1989), 76.

7 Carol J. Nicholson, 'Elegance and Grass Roots: The Neglected Philosophy of Frederick Law Olmsted', *Transactions of the Charles S. Peirce Society* 40, no. 2 (2004): 335–48.

8 John Ruskin is a fine example of an advocate for nature in the nineteenth century. See John Ruskin and Clive Wilmer (eds), *Unto this Last: and Other Writings* (London: Penguin, 1985).

9 Richard Balding, 'Turn Off Your Smartphone to Beat Stress'. *British Psychological Society*, 13 June (2012). Available online: http://wp.me/p2tvLx-cz (accessed 11 June 2017).

10 Alan Mezes, 'Your Smartphone May be Stressing you Out'. *USA Today*, 13 January (2012). Available online: http://usatoday30.usatoday.com/news/health/story/health/story/2012-01-13/Your-smartphone-may-be-stressing-you-out/52529514/1 (accessed 11 June 2017).

11 Raymond S. Nickerson, 'Confirmation Bias: A Ubiquitous Phenomenon in Many Guises', *Review of General Psychology* 2, no. 2 (1998): 175–220.

12 Richard Coyne, *Technoromanticism: Digital Narrative, Holism, and the Romance of the Real* (Cambridge, MA: MIT Press, 1999).

13 For a critical account of 'the new organic architecture', see Philip Steadman, *Evolution of Designs: Biological Analogy in Architecture and the Applied Arts* (London: Routledge, 2008), 238.

14 Olga Budashevskaya and Blaine Brownell, *Hypernatural: Architecture's New Relationship with Nature* (Princeton, NJ: Princeton University Press, 2015), 21.

15 Ibid.

16 See Dennis Dollens, 'A System of Digital-botanic Architecture', *Leonardo* 38, no. 1 (2005): 15–21, Stephen R. Kellert, et al., *Biophilic Design: The Theory, Science and Practice of Bringing Buildings to Life* (Hoboken, NJ: Wiley, 2008); Sue Thomas, *Technobiophilia: Nature and Cyberspace* (London: Bloomsbury, 2013), Dennis Dollens, *Autopoietic-extended Architecture: Can Buildings Think? (PhD Thesis)* (Edinburgh: The University of Edinburgh, 2015).

17 Steadman, *Evolution of Designs: Biological Analogy in Architecture and the Applied Arts*

18 D'Arcy Wentworth Thompson, *On Growth and Form* (Cambridge, England: Cambridge University Press, 2014).

19 Architectural theorist Chris Abel also makes this point clearly. See Chris Abel, *The Extended Self: Architecture, Memes and Minds* (Manchester, England: Manchester University Press, 2015), 150.

20 There are other models of design, many antithetical to the evolutionary approach. See, for example, Richard Coyne, *Designing Information Technology in the Postmodern Age: From Method to Metaphor* (Cambridge, MA: MIT Press, 1995); Bill Gaver, et al., 'Design: Cultural Probes', *Interactions Magazine* 6, no. 1 (1999): 21–9; Richard Buchanan, 'Children of the Moving Present: The Ecology of Culture and the Search for Causes in Design', *Design Issues* 17, no. 1 (2001): 67–84; Adrian Snodgrass and Richard Coyne, *Interpretation in Architecture: Design as a Way of Thinking* (London: Routledge, 2006).

21 John Chris Jones, *Design Methods: Seeds of Human Futures* (London: Wiley, 1970).

22 For a summary, see Harry Francis Mallgrave and David Goodman, *An Introduction to Architectural Theory: 1968 to the Present* (Chichester, England: Wiley-Blackwell, 2011).

23 George Bull, *Public Health and Landscape: Creating Healthy Places* (London: Landscape Institute, 2013), 1.

24 See the RIBA website: https://www.architecture.com/RIBA/Aboutus/Whoweare/Whoweare.aspx (accessed 3 June 2017).

25 Martin Heidegger, *Being and Time* (London: SCM Press, 1962), 100.

26 Charles Jencks and George Baird (eds), *Meaning in Architecture* (London: Barrie & Rockliff, 1969). In a similar vein, the issue of meaning in landscape architecture was taken up by Marc Treib, 'Must Landscapes Mean?', in *Theory in Landscape Architecture: A Reader*, ed. S. Swaffield (Philadelphia, PA: University of Pennsylvania Press), 89–102; Jane Gillette, 'Can Gardens Mean?', *Landscape Journal* 24, no. 1 (2005): 85–97; Marc (ed.), Treib, *Meaning in Landscape Architecture and Gardens: Four Essays, Four Commentaries* (London: Abingdon, Oxon, 2011).

27 Charles Sanders Peirce, *The Collected Papers of Charles Sanders Peirce. Vol. 2 Elements of Logic* (Cambridge, MA: Harvard University Press, 1932); Charles Sanders Peirce, *The Collected Papers of Charles Sanders Peirce. Vol. 5 Pragmatism and Pragmaticism* (Cambridge,

MA: Harvard University Press, 1935); Charles Sanders Peirce, *The Essential Peirce, Selected Philosophical Writings Volume 2 (1893-1913)* (Bloomington, IN: Indiana University Press, 1998).

28 Ferdinand de Saussure, *Course in General Linguistics* (London: Duckworth, 1983); Fredric Jameson, *The Prison-House of Language: A Critical Account of Structuralism and Russian Formalism* (Princeton, NJ: Princeton University Press, 1972); Terrence Hawkes, *Structuralism and Semiotics* (London: Routledge, 2003).

29 Thomas A. Sebeok, *Signs: An Introduction to Semiotics* (Toronto: University of Toronto Press, 1999).

30 Marcello (ed.) Barbieri, *Introduction to Biosemiotics: The New Biological Synthesis* (Dordrecht, the Netherlands: Springer, 2008).

31 Bronislaw Szerszynski, 'The End of the End of Nature: The Anthropocene and the Fate of the Human', *Oxford Literary Review* 34, no. 2 (2012): 165–84.

32 Snodgrass and Coyne, *Interpretation in Architecture: Design as a Way of Thinking.*

33 Richard Coyne, *Derrida for Architects* (Abingdon: Routledge, 2011).

34 Richard Coyne, *The Tuning of Place: Sociable Spaces and Pervasive Digital Media* (Cambridge, MA: MIT Press, 2010); Richard Coyne, *Mood and Mobility: Navigating the Emotional Spaces of Digital Social Networks* (Cambridge, MA: MIT Press, 2016).

35 William James, *The Principles of Psychology Volume I* (Cambridge, MA: Harvard University Press, 1981).

36 Nathan Houser, 'Introduction', in *The Essential Peirce, Selected Philosophical Writings Volume 2 (1893-1913)*, ed. N. Houser (Bloomington, IN: Indiana University Press), xvii–xxxviii, xxxii.

37 Thoreau, *Walden: Or, Life in the Woods.*

38 Panagiotis Mavros, et al., 'Engaging the Brain: Implications of Mobile EEG for Spatial Representation', in *Digital Physicality Proceedings of the 30th eCAADe Conference*, eds H. Achten, J. Pavlicek, J. Hulin and D. Matejdan (Czech Technical University in Prague: Molab), 657–65; Peter Aspinall, et al., 'The Urban Brain: Analysing Outdoor Physical Activity with Mobile EEG', *British Journal of Sports Medicine* 49, no. 4 (2013): http://dx.doi.org/10.1136/bjsports-2012-091877; Jenny J. Roe, et al., 'Engaging the Brain: The Impact of Natural Versus Urban Scenes using Novel EEG Methods in an Experimental Setting', *Environmental Sciences* 1, no. 2 (2013): 93–104.

39 See my blog site at https://richardcoyne.com (accessed 12 June 2017).

Chapter 1

1 Frequency band for British Broadcasting Corporation (BBC) radio 4.

2 Saussure, *Course in General Linguistics.*

3 C. K. Ogden and I. A. Richards, *The Meaning of Meaning: A Study of the Influence of Language upon Thought and of the Science of Symbolism* (Orlando, FL: Harvest, 1989).

4 Roman Jakobson and Morris Halle, *Fundamentals of Language* (The Hague, Holland: Mouton, 1956).

5 Charles Sanders Peirce, *The Essential Peirce, Selected Philosophical Writings Volume 1 (1867-1893)* (Bloomington, IN: Indiana University Press, 1992); Peirce, *The Essential Peirce, Selected Philosophical Writings Volume 2 (1893–1913).*

6 W. James, *The Principles of Psychology* (New York: Dover, 1950).

7 John Dewey, *Experience and Nature* (New York: Dover, 1958).

8 Houser, 'Introduction', xxxii.

9 On *attunement* in spatial experience, see my book Coyne, *The Tuning of Place: Sociable Spaces and Pervasive Digital Media* and also Alberto Perez-Gomez, *Attunement: Architectural Meaning After the Crisis of Modern Science* (Cambridge, MA: MIT Press, 2016).

10 Henri Lefebvre, *Rhythmanalysis: Space, Time and Everyday Life* (London: Continuum, 2004).

11 Tristan Gooley, *The Walker's Guide to Outdoor Clues and Signs* (London: Sceptre, Hodder and Stoughton, 2014), 106.

12 Martin Heidegger deals with the issue of attention as the practical experience of the 'ready-to-hand' (zuhanden). See Heidegger, *Being and Time*, 98.

13 I will address the idea of our potential fascination with nature in Chapter 9. See Stephen Kaplan, 'The Restorative Benefits of Nature: Toward an Integrative Framework', *Journal of Environmental Psychology* 15, no. (1995): 169–82. For an account of how our attention is 'administered' in the 'attention economy' see Deborah Hauptmann, 'Introduction: Architecture and Mind in the Age of Communication and Information', in *Cognitive Architecture: From Biopolitics to Noopolotics. Architecture and Mind in the Age of Communication and Information*, ed. D. Hauptmann and W. Neidich (Rotterdam: 010 Publishers), 10–45, 24–5.

14 James, *The Principles of Psychology Volume I*, 285.

15 We can think here of disgust responses. See Mary Douglas, *Purity and Danger: An Analysis of Concepts of Pollution and Taboo* (London: Routledge and Kegan Paul, 1966).

16 James, *The Principles of Psychology Volume I*, 395.

17 The project is called 'Mobility, Mood and Place: a user-centred approach to design of built environments to make mobility easy, enjoyable and meaningful for older people' and is supported by the EPSRC/AHRC/SRC/MRC scheme Design for Well-being, Ageing and Mobility in the Built Environment (EP/K037404).

18 Daniel J. Simons and Christopher F. Chabris, 'Gorillas in Our Midst: Sustained Inattentional Blindness for Dynamic Events', *Perception* 28, no. (1999): 1059–74.

19 James, *The Principles of Psychology Volume I*, 411.

20 As examined by Michel Foucault, *Discipline and Punish: The Birth of the Prison* (London: Penguin, 1977).

21 See Nel Noddings, *Caring: A Feminine Approach to Ethics and Moral Education* (Berkeley, CA: University of California Press, 1984) Quoted in Charles Guignon, *The Good Life* (Indianapolis, IL: Hackett, 1999), 320.

22 See an account of empathy by Adam Smith, *The Theory of Moral Sentiments* (Indianapolis: Liberty Fund, 1984).

23 James, *The Principles of Psychology Volume I*, 411.

24 Geoffrey Broadbent, *Design in Architecture: Architecture and the Human Sciences* (New York, NY: John Wiley and Son, 1973).

25 Donald Preziosi, *Architecture, Language and Meaning: Origins of the Built World and Its Semiotic Organization* (The Hague, The Netherlands: Mouton, 1979).

26 Michel Foucault, *The Order of Things: An Archaeology of the Human Sciences* (New York: Random House, 1970).

27 Wilhelm Wundt, *Outlines of Psychology* (Leipzig: W. Engelmann, 1897).

28 Also see James A. Russell and Geraldine Pratt, 'A Description of the Affective Quality Attributed to Environments', *Journal of Personality and Social Psychology* 38, no. 2 (1980): 311–22.

29 Mog Stapleton, 'Feeling the Strain: Predicting the Third Dimension of Core Affect', *Behavioural and Brain Sciences* 35, no. 3 (2012): 166–7.

30 See John Dewey, *Art as Experience* (New York: Wideview Perigee, 1980), 16.

31 Edmund Husserl, *Ideas: General Introduction to Pure Phenomenology* (New York: Collier Books, 1962), 93.

32 Ibid.

33 Contrary to Husserl, Heidegger emphasizes this attending as a break from our practical, unreflective engagement in our everyday 'concernful dealings' in the world, which Hubert Dreyfus in his commentary on *Being and Time* describes as 'a disturbance'. Hubert L. Dreyfus, *Being-in-the-world: A Commentary on Heidegger's Being and Time, Division I* (Cambridge, MA: MIT Press, 1991), 70.

34 Claude E. Shannon and William Weaver, *The Mathematical Theory of Communication* (Urbana, IL: The University of Illinois Press, 1963).

35 For example, see Michael Reddy, 'The Conduit Metaphor: A Case of Frame Conflict in Our Language about Language', in *Metaphor and Thought*, ed A. Ortony (Cambridge: Cambridge University Press), 284–324.

36 Charles Sanders Peirce, 'What Makes a Reasoning Sound?', in *The Essential Peirce, Selected Philosophical Writings Volume 2 (1893-1913)*, ed. N. Houser (Bloomington, IN: Indiana University Press), 242–57, 256.

37 Ezra M. Markowitz and Azim F. Shariff, 'Climate Change and Moral Judgement', *Nature Climate Change* 2, no. (2012): 243–7, 244.

38 Ibid.

39 On intentions and agency, also see Roland Barthes, 'The Death of the Author', in *Image, Music, Text*, ed. S. Heath (London: Fontana), 142–9.

40 Husserl, *Ideas: General Introduction to Pure Phenomenology.*

41 Ibid., 106.

42 Heidegger, *Being and Time*. In comparing phenomenology with empirical psychology, P. Sven Arvidson says that in Husserl's opinion 'intentionality is fundamental and attentionality is an extremely important modification of it' (203). Husserl does not speak favourably of semiotics. See Göran Sonesson, 'Phenomenology Meets Semiotics: Two not so Very Strange Bedfellows at the End of their Cinderella Sleep', *Metodo. International Studies in Phenomenology and Philosophy* 3, no. 1 (2015): 41–62.

43 W. J. T. Mitchell, 'Imperial Landscape', in *Landscape and Power*, ed. W. J. T. Mitchell (Chicago: University of Chicago Press), 5–34.

44 Gooley, *The Walker's Guide to Outdoor Clues and Signs*, 10.

45 Jay Appleton, *The Experience of Landscape* (London: John Wiley, 1975) For an interesting critique of Appleton's position on prospect and refuge, see W. J. T. Mitchell, *Landscape and Power* (Chicago: University of Chicago Press, 2002), 16.

46 Jay Appleton, 'Prospects and Refuges Re-visited', *Landscape Journal* 3, no. 2 (1984): 91–103, 102.

47 Gordon Cullen, *Concise Townscape* (Abingdon, England: Architectural Press, 1961).

48 Ibid., 17.

49 Ibid., 19.

50 Donald Appleyard, et al., *The View from the Road* (Cambridge, MA: Joint Center for Urban Studies of the Massachusetts Institute of Technology and Harvard University, 1966), 6.

51 Ibid.

52 Ibid., 36.

53 Sylvia Crowe, *The Landscape of Power* (London: Architectural Press, 1958). For an example of how views affect journeys through landscapes, see Catharine Ward Thompson, 'Landscape Quality and Quality of Life', in *Innovative Approaches for Researching Landscape and Health: Open Space: People Space 2*, ed. C. Ward Thompson, P. Aspinall and S. Bell (Abingdon, England: Routledge), 230–55.

54 Appleyard, et al., *The View from the Road,* 33.

55 Evelyn Waugh, *Brideshead Revisited: The Sacred and Profane Memories of Captain Charles Ryder* (Harmondsworth, England: Penguin, 1951), 43.

56 C. S. Lewis, *The Lion, the Witch and the Wardrobe* (London: Geoffrey Bles, 1950).

57 For an experiment simulating such moments of transition using video images, see Harry Heft and Jack L. Nasar, 'Evaluating Environmental Scenes Using Dynamic Versus Static Displays', *Environment and Behaviour* 32, no. 3 (2000): 301–22.

58 From Lucretius, *On the Nature of the Universe*, trans R. E. Latham (London: Penguin, 1994). Quoted in Guignon, *The Good Life,* 43.

59 From Blaise Pascal, *Pensées*, trans A. J. Krailsheimer (London: Penguin, 1995). Quoted in ibid., 199.

60 Sascha Topolinski and Rolf Reber, 'Gaining Insight into the "Aha" Experience', *Current Directions in Psychological Science* 19, no. 6 (2010): 402–5.

61 Coyne, *The Tuning of Place: Sociable Spaces and Pervasive Digital Media.*

62 For example, see Daniel Cernea, et al., 'Detecting Insight and Emotion in Visualization Applications with a Commercial EEG Headset', in *Proc. SIGRAD 2011*, 53–60.

63 Richard Coyne (2016), 'Brainwalks'. *Reflections on Technology, Media and Culture*, 18 March. Available online: https://richardcoyne.com/2017/03/18/brainwalks/ (accessed 14 June 2017).

64 See https://www.emotiv.com (accessed 14 June 2017).

65 Aspinall, et al., 'The Urban Brain: Analysing Outdoor Physical Activity with Mobile EEG'.

66 On the psychology frustration, see Sigmund Freud, 'Infantile Sexuality', in *The Penguin Freud Library, Volume 7: On Sexuality*, ed A. Richards (Harmondsworth, Middlesex: Penguin), 88–126.

67 Marc Augé, *Non-places: Introduction to an Anthropology of Supermodernity* (London: Verso, 1995).

68 James, *The Principles of Psychology Volume I,* 121.

69 Coyne, *The Tuning of Place: Sociable Spaces and Pervasive Digital Media.*

70 James, *The Principles of Psychology Volume I,* 114.

71 Ibid., 104.

72 Dewey, *Experience and Nature,* 3a.

73 See Bratton, *The Stack: On Software and Sovereignty*, 137. Also see Carole Cadwalladr, (2017), 'Robert Mercer: The Big Data Billionaire Waging War on Mainstream Media'. *Guardian*, 26 February. Available online: https://www.theguardian.com/politics/2017/feb/26/robert-mercer-breitbart-war-on-media-steve-bannon-donald-trump-nigel-farage (accessed 11 June 2017).

74 Vance Packard, *The Hidden Persuaders* (Brooklyn, NY: Ig Publishing, 2007).

75 Sherry Turkle, *Alone Together: Why We Expect More from Technology and Less from Each other* (New York: Basic Books, 2011), The New State of the Self: Tethered and Marked Absent.

Chapter 2

1 Martin Heidegger, *The Question Concerning Technology and Other Essays* (New York: Harper and Row, 1977).

2 Thomas P. Hughes, 'The Evolution of Large Technological Systems', in *The Social Construction of Technological Systems: New Directions in the Sociology and History of Technology*, ed W. E. Bijker, T. P. Hughes, T. Pinch and D. G. Douglas (Cambridge, MA: MIT Press), 45–76.

3 Coyne, *Technoromanticism: Digital Narrative, Holism, and the Romance of the Real*; Coyne, *Designing Information Technology in the Postmodern Age: From Method to Metaphor*; Richard Coyne, *Cornucopia Limited: Design and Dissent on the Internet* (Cambridge, MA: MIT Press, 2005); Coyne, *The Tuning of Place: Sociable Spaces and Pervasive Digital Media*; Coyne, *Mood and Mobility: Navigating the Emotional Spaces of Digital Social Networks.*

4 Fred Turner, *From Counterculture to Cyberculture: Stewart Brand, the Whole Earth Network, and the Rise of Digital Utopianism* (Chicago, IL: The University of Chicago Press, 2006); Fred Turner, *The Democratic Surround: Multimedia and American Liberalism from World War II to the Psychedelic Sixties* (Chicago, IL: The University of Chicago Press, 2013).

5 Butler, *Gender Trouble: Feminism and the Subversion of Identity*, 7.

6 Howard Rheingold, *The Virtual Community: Homesteading on the Electronic Frontier* (Reading, MA: Addison-Wesley, 1993); Walter Isaacson, *Steve Jobs: The Exclusive Biography* (London: Hachette Digital, 2011).

7 Koert van Mensvoort and Hendrik-Jan Grievink, *Next Nature: Nature Changes Along With Us* (Barcelona, Spain: Actar, 2011).

8 Thoreau, *Walden: Or, Life in the Woods*, 89.

9 As I explored in Coyne, *The Tuning of Place: Sociable Spaces and Pervasive Digital Media.*

10 Peirce, 'Immortality in the Light of Synechism', 256. He continues 'in communicating our knowledge to others and in getting at the knowledge that those others seek to communicate to us'. He also says, 'the meaning of a symbol consists in *how* it might cause us to act' (p. 202).

11 Phil Macnaghten and John Urry, *Contested Natures* (London: Sage, 1998), 1.

12 Christian Norberg-Schulz, *Genius Loci: Towards a Phenomenology of Architecture* (New York: Rizzoli, 1980).

13 Macnaghten and Urry, *Contested Natures*, 1.

14 Ibid., 140.

15 Ibid. It is worth noting a niche trend to explore dark, polluted places. See Andrew Blackwell, *Visit Sunny Chernobyl: Adventures in the World's Most Polluted Places* (London: Random, 2013).

16 Steadman, *Evolution of Designs: Biological Analogy in Architecture and the Applied Arts*, 4.

17 Ibid., 239.

18 Ibid., 240.

19 Perez-Gomez, *Attunement: Architectural Meaning After the Crisis of Modern Science*, 2–3.

20 See Charles Jencks, *Modern Movements in Architecture* (New York: Penguin, 1973); Kenneth Frampton, *Modern Architecture: A Critical History* (London: Thames and Hudson, 1985); Nikolaus Pevsner, *Pioneers of Modern Design: From William Morris to Walter Gropius* (New Haven, CN: Yale University Press, 2005); Iain Boyd Whyte (ed.), *Modernism and the Spirit of the City* (London: Routledge, 2003). Le Corbusier advocated for an architecture that returns people to nature. See Emma Dummett, *Green Space and Cosmic Order: Le Corbusier's Understanding of Nature* (Edinburgh: PhD Thesis, The University of Edinburgh, 2007).

21 Oliver Botar, 'The Biocentric Bauhaus', in *The Routledge Companion to Biology in Art and Architecture*, ed. C. N. Terranova and M. Tromble (London: Routledge), 17–51, 36.

22 Ibid.

23 L. R. Furst, *Romanticism in Perspective: A Comparative Study of Aspects of the Romantic Movements in England, France and Germany* (London: MacMillan, 1969). For a discussion of the city and the unconscious, and the departure of structuralism from the biologism of the nineteenth century, see John Shannon Hendrix and Lorens Eyan Holm, 'Aldo Rossi and the Field of the Other', in *Architecture and the Unconscious*, ed. J. S. Hendrix and L. E. Holm (Farnham: Ashgate/Routledge), 99–117.

24 Botar, 'The Biocentric Bauhaus', 17.

25 Ibid., 22.

26 Ibid., 23.

27 Raoul Heinrich Francé, *Die Pflanze als Erfinder (The Plant as an Inventor)* (Kosmos: Stuttgart, 1920).

28 Botar, 'The Biocentric Bauhaus', 20.

29 Bauhaus teacher László Moholy-Nagy said of the studio teacher: 'The teacher who has come to a full realization of the organic oneness and the harmonious sense of rhythm of life should have a tongue of fire to expound his happiness.' Sibyl Moholy-Nagy, *Moholy-Nagy: Experiment in Totality* (Cambridge, MA: MIT Press, 1950), 44–5.

30 Francé, *Die Pflanze als Erfinder (The Plant as an Inventor)* See English translation at http://www.joostrekveld.net/?p=574 (accessed 3 June 2017).

31 Botar, 'The Biocentric Bauhaus', 22.

32 Marcus Aurelius, *Meditations* (London: Penguin, 1964), 73.

33 For an account of cities as 'netspaces', see Katharine S. Willis, *Netspaces: Space and Place in a Networked World* (Farnham, UK: Ashgate, 2016). Also see Stephen Graham and Simon Marvin, *Splintering Urbanism: Networked Infrastructures, Technological Mobilities and the Urban Condition* (London: Routledge, 2001); William J. Mitchell, *Me++: The Cyborg Self and the Networked City* (Cambridge, MA: MIT Press, 2003); Bill Hillier, 'The City as Socio-technical System: A Spatial Reformulation in the Light of the Levels Problem and the Parallel Problem', in *Digital Urban Modelling and Simulation*, ed. S. Müller Arisona, A.

Gideon, J. Halatsch and P. Wonka (Heidelberg, Germany: Springer), 24–48. On network society in general, see Starr Roxanne Hiltz and Murray Turoff, *The Network Nation: Human Communication via Computer* (Cambridge, MA: MIT Press, 1993); Manuel Castells, *The Rise of the Network Society* (Malden, MA: Blackwell Publishers, 1996). For an architectural critique of networks, see Mark Wigley, 'Network Fever', *Grey Room* 4, no. (2001): 82–122, and, for my own review, see Richard Coyne, 'The Net Effect: Design, the Rhizome, and Complex Philosophy', *Futures* 40, no. (2008): 552–61.

34 As dictated by the Renaissance architect Leon Battista Alberti. See Leon Battista Alberti, *On the Art of Building in Ten Books* (Cambridge, MA: MIT Press, 1996).

35 James Gleick, *Chaos: Making a New Science* (London: Heinemann, 1988); David Easley and Jon Kleinberg, *Networks, Crowds and Markets: Reasoning About a Highly Connected World* (Cambridge, UK: Cambridge University Press, 2010).

36 See Manuel Castells, et al., *Mobile Communication and Society: A Global Perspective* (Cambridge, MA: MIT Press, 2007).

37 Saket Navlakha and Ziv Bar-Joseph, 'Distributed Information Processing in Biological Systems', *Communication of the ACM* 58, no. 1 (2015): 94–102, 101.

38 Donna J. Haraway, *Simians, Cyborgs, and Women: The Reinvention of Nature* (London: FAb, 1991), 161.

39 Ibid., 170.

40 Ibid., 170. Networking as a form of weaving is interesting. See Coyne, 'The Net Effect: Design, the Rhizome, and Complex Philosophy'.

41 See Sadie Plant, *Zeros and Ones: Digital Women and the New Technoculture* (London: Fourth Estate, 1998); Coyne, 'The Net Effect: Design, the Rhizome, and Complex Philosophy'.

42 Navlakha and Bar-Joseph, 'Distributed Information Processing in Biological Systems', 101.

43 Ibid., 98.

44 Brittany Ransom, 'The Sixth Element: DIY Cyborgs and the Hive Minds of Social Media', in *The Routledge Companion to Biology in Art and Architecture*, ed. C. N. Terranova and M. Tromble (London: Routledge), 458–66, 461.

45 Ibid.

46 Angus Macdonald, *Structure in Architecture* (Abingdon, Oxon: Taylor and Francis, 2001).

47 Geoffrey Broadbent. 'Building design as an iconic sign system', in *Signs, Symbols, and Architecture*, ed G. Broadbent, R. Bunt and C. Jencks, (Chichester, England: John Wiley and Sons), 311-31, 124.

48 Preziosi, *Architecture, Language and Meaning: Origins of the Built World and Its Semiotic Organization*, 4.

49 Ibid.

50 Ibid.

51 Ibid.

52 Ibid.

53 Ibid., 5.

54 Ibid., 67.

55 Richard Coyne, et al., *Knowledge-Based Design Systems* (Reading, Mass.: Addison-Wesley, 1990).

56 Andreas Papadakis, et al., *Deconstruction: Omnibus Volume* (London: Academy Editions, 1989); Geoffrey Broadbent and Jorge Glusberg (ed.), *Deconstruction: A Student Guide* (London: 1991); Coyne, *Derrida for Architects.*

57 See, for example, Jonathan Hale, *Merleau-Ponty for Architects* (London: Routledge, 2016).

58 Preziosi, *Architecture, Language and Meaning: Origins of the Built World and Its Semiotic Organization,* 28.

59 Ibid.

60 Ibid., 39.

61 Jencks and Baird (eds), *Meaning in Architecture.*

62 Charles Jencks, *The Architecture of the Jumping Universe, A Polemic: How Complexity Science Is Changing Architecture and Culture* (London: Academy Editions, 1995).

63 Ibid., 125.

64 Ibid., 48.

65 Ibid.

66 Alternatively, I could say that representation is never mere representation, and signs are never just signs.

67 John Palmesino, et al., 'Matters of observation on architecture in the Anthropocene', in *Architecture in the Anthropocene: Encounters Among Design, Deep Time, Science and Philosophy*, ed E. Turpin, (Ann Arbor, MI: Open Humanities Press), 15-24, 20

68 Coyne, *Designing Information Technology in the Postmodern Age: From Method to Metaphor*

69 John McCarthy and Peter Wright, *Technology as Experience* (Cambridge, MA: MIT Press, 2004)

70 Peirce, 'The maxim of Pragmatism'.

71 Ibid., 135.

72 Heidegger, *Being and Time*

73 Dreyfus, *Being-in-the-world: A Commentary on Heidegger's Being and Time, Division I,* 6

74 Coyne, *Derrida for Architects.*

Chapter 3

1 Sebeok, *Signs: An Introduction to Semiotics,* 26.

2 Dewey, *Experience and Nature,* i.

3 Harvey J. Miller and Michael F. Goodchild, 'Data-driven Geography', *GeoJournal* 80, no. (2015): 449–61, 449.

4 Edd Dumbill (2012), 'What is Big Data? An Introduction to the Big Data Landscape'. *O'Reilly,* 11 January. Available online: https://beta.oreilly.com/ideas/what-is-big-data (accessed 11 June 2017).

5 For an account of 'the cloud' as a political, economic and social enterprise, see Bratton, *The Stack: On Software and Sovereignty.*

6 Google's search engine has a reverse image-recognition system. You upload an image and Google returns similar images from its extensive (big data) database, using image recognition,

machine learning and neural network techniques. See Chuck Rosenberg (2013), 'Improving Photo Search: A Step Across the Semantic Gap'. *Google Research Blog*, 12 June. Available online: https://research.googleblog.com/2013/06/improving-photo-search-step-across.html (accessed 3 June 2017).

7 Tony Hey, et al., *The Fourth Paradigm: Data-Intensive Scientific Discovery* (Redmond, WA: Microsoft Research, 2009).

8 Chris Anderson (2008), 'The End of Theory: The Data Deluge Makes the Scientific Method Obsolete'. *Wired Magazine*, 23 June. Available online: http://archive.wired.com/science/discoveries/magazine/16-07/pb_theory (accessed 11 June 2017).

9 For a critique of big-data narratives and the smart city, see Alex Aurigi, 'No Need to Fix: Strategic Inclusivity in Developing and Managing the Smart City', in *Digital Futures and the City of Today: New Technologies and Physical Spaces*, ed. G. A. Caldwell, S. C. H. and E. M. Clift (Bristol: Intellect), 9–27.

10 Victor Mayer-Schonberger and Kenneth Cukier, *Big Data: A Revolution That Will Transform How We Live, Work, and Think* (Boston, MA: Eamon Dolan, 2013), 193.

11 Ibid., 192.

12 Ibid., 193.

13 Ibid. Hence the idea of the secure distributed data ledger and its problems. See 'Blockchains: The Great Chain of Being Sure about Things' (2015), *The Economist*, 31 October. Available online: http://www.economist.com/news/briefing/21677228-technology-behind-bitcoin-lets-people-who-do-not-know-or-trust-each-other-build-dependable (accessed 15 June 2017).

14 Mayer-Schonberger and Cukier, *Big Data: A Revolution That Will Transform How We Live, Work, and Think*, 193.

15 See Christopher T. Stout and Reuben Kline, 'Racial Salience, Viability, and the Wilder Effect: Evaluating Polling Accuracy for Black Candidates', *Public Opinion Quarterly* 79, no. 4 (2015): 994–1014. Note the so-called 'Hawthorne effect', where workers adjust their behaviour once they know they are part of an observation study. See F. J. Roethlisberger and W. J. Dickson, *Management and the Worker* (Cambridge, MA: Harvard University Press, 1939).

16 Dumbill, 'What is Big Data? An Introduction to the Big Data Landscape'.

17 Adam Greenfield and Mark Shepard, *Situated Technologies Pamphlets 1: Urban Computing and its Discontents* (New York: The Architectural League of New York, 2007); Adam Greenfield and Kim Nurri, *Against the Smart City (The city is here for you to use Book 1)* (New York, NY: Do Projects, 2013) Antoine Picon, *Smart Cities: A Spatialised Intelligence* (London: Wiley, 2015).

18 Rob Kitchin, 'The Real-time City? Big Data and Smart Urbanism', *GeoJournal* 79, no. (2014): 1–14, 2.

19 Ibid.

20 Franco Moretti, *Distant Reading* (London: Verso, 2013).

21 Ibid., 299.

22 Kathryn Schulz (2011), 'What is Distant Reading?'. *The New York Times*, 24 June. Available online: http://www.nytimes.com/2011/06/26/books/review/the-mechanic-muse-what-is-distant-reading.html?_r=0 (accessed 11 June 2017).

23 Tricia Wang (2013), 'Big Data Needs Thick Data'. *Ethnography Matters*, 13 May. Available online: http://ethnographymatters.net/blog/2013/05/13/big-data-needs-thick-data/ (accessed 11 June 2017).

24 Anderson, 'The End of Theory: The Data Deluge Makes the Scientific Method Obsolete'.

25 Michael Gillam, et al., 'The Healthcare Singularity and the Age of Semantic Medicine', in *The Fourth Paradigm: Data-Intensive Scientific Discovery*, ed. T. Hey, S. Tansley and K. Tolle (Redmond, WA: Microsoft Research), 57–64, 61.

26 Ibid.

27 Ibid.

28 Dumbill, 'What is Big Data? An Introduction to the Big Data Landscape'.

29 Ibid.

30 An idea that was given twentieth-century treatment by Roger Penrose. See Roger Penrose, *The Emporer's New Mind: Concerning Computers, Minds, and the Laws of Physics* (London: Vintage, 1989).

31 Heidegger, *The Question Concerning Technology and Other Essays.*

32 Lily E. Kay, *Who Wrote the Book of Life? A History of the Genetic Code* (Stanford: Stanford University Press, 2000). For an extended comparison between approaches to biology from semiotics and information theory, see Gérard Battail, 'Applying Semiotics and Information Theory to Biology: A Critical Comparison', *Biosemiotics* 2, no. (2009): 303–20.

33 Peirce, *The Collected Papers of Charles Sanders Peirce. Vol. 2 Elements of Logic.*

34 Sebeok, *Signs: An Introduction to Semiotics.*

35 Ibid., 28.

36 Ibid., 28–9.

37 Semiotics also asserts that language is but a subspecies of semiotics. According to psychologist Jean Piaget, 'language is merely one particular instance of the semiotic or symbolic function' (Jean Piaget, *Biology and Knowledge* (Edinburgh, UK: Edinburgh University Press, 1971), 46). Semiotics taps into systems of imitation, gestures, play, mental imagery, drawing and other actions that derive from sensorimotor experience.

38 Sebeok, *Signs: An Introduction to Semiotics,* 46.

39 Ibid., 66.

40 Ogden and Richards, *The Meaning of Meaning: A Study of the Influence of Language upon Thought and of the Science of Symbolism,* 21.

41 Sebeok, *Signs: An Introduction to Semiotics,* 10.

42 Ibid., 52.

43 Ibid., 10.

44 Ibid., 52.

45 Ibid., 84. This reminds me of Roger Caillois's 1935 article about insect mimicry, that he sees as a spatial disturbance. See Roger Caillois, 'Mimicry and Legendary Psychasthenia', *October* 31, no. Winter (1984): 17–32 (First published in *Minotaure* in 1935).

46 Sebeok, *Signs: An Introduction to Semiotics,* 87. He is quoting from Peirce, *The Essential Peirce, Selected Philosophical Writings Volume 2 (1893–1913),* 291.

47 Sebeok, *Signs: An Introduction to Semiotics,* 89.

48 Ibid., 90–1.

49 Albert Atkin, 'Peirce on the Index and Indexical Reference', *Transaction of the Charles S. Peirce Society* 41, no. 1 (2005): 161–88, 165.

50 Sebeok, *Signs: An Introduction to Semiotics,* 39.

51 Ibid., 94.

52 Ibid., 96.

53 Ibid., 58.

54 Ibid., 60.

55 Ibid., 105.

56 Jennifer Rowley, 'The Wisdom Hierarchy: Representations of the DIKW Hierarchy', *Journal of Information Science* 33, no. 2 (2007): 163–80.

57 Steven Levy, *Artificial Life: The Quest for a New Creation* (London, England: Penguin, 1992); Katherine N. Hayles, 'Narratives of Artificial Life', in *Future Natural: Nature, Science, Culture*, ed. G. Robertson, et al. (London: Routledge), 146–64.

58 Hans P. Moravec, *Mind Children: The Future of Robot and Human Intelligence* (Cambridge, MA: Harvard University Press, 1988).

59 Eduardo Kohn, *How Forests Think: Anthropology Beyond the Human* (Berkeley, CA: University of California Press, 2013).

60 Ibid., 74.

61 Ibid.

62 There are several versions of this story. In one cautionary tale, two children are lured into the forest where they die and their corpses are covered over by robins. Randolph Caldecott (2006), 'The Babes in the Wood'. Project Gutenberg EBook. Available online: http://www.gutenberg.org/files/19361/19361-h/19361-h.htm (accessed 11 June 2017).

63 J. R. R. Tolkien, *The Lord of the Rings* (London: Harper Collins, 2005).

64 Robert A. Wilson and Andy Clark. 'How to Situate Cognition: Letting Nature Take its Course', in *The Cambridge Handbook of Situated Cognition*, ed. M. Aydede and P. Robbins (Cambridge, England: Cambridge University Press), 5, 55–77.

65 Ibid., 6.

66 Ibid., 11.

67 Ibid., 12.

68 For an account of the situated nature of cognition, see Edwin Hutchins, *Cognition in the Wild* (Cambridge, MA: MIT Press, 1996).

69 Rather than 'world of experience', Heidegger would say more carefully 'mode of being', but I think the approximation here is sufficient.

70 This is the 'pure occurrent'. See Dreyfus, *Being-in-the-world: A Commentary on Heidegger's Being and Time, Division I*, 125 , and Heidegger, *Being and Time.*

71 Anderson, 'The End of Theory: The Data Deluge Makes the Scientific Method Obsolete'.

72 Dreyfus, *Being-in-the-world: A Commentary on Heidegger's Being and Time, Division I*, 100–2.

Chapter 4

1 Alessandro Delfanti, *Biohackers: The Politics of Open Science* (London: PlutoPress, 2013), 111.

2 Ibid., 60.

3 Ibid., 113.

4 Ibid., 115. Such disruptive experimentation is on a par with 'uberization' (named after Uber, the car-hire company): a change in market services introduced via a new, distributed model for which there is as yet no regulation.

5 Ibid., 127.

6 https://www.23andme.com/en-gb/health/ (accessed 8 February 2017).

7 Delfanti, *Biohackers: The Politics of Open Science*, 2.

8 Ibid., 74.

9 Christopher Langton (1989), 'Artificial Life'. Available online: http://90.146.8.18/de/archiv_files/19931/1993_025.pdf (accessed 3 June 2017).

10 Ibid. Also see Christopher G. Langton (ed.), *Artifical Life: An Overview* (Cambridge, MA: MIT Press, 1997).

11 William T. Keeton, *Biological Science* (New York: Norton, 1972), 521.

12 Richard Dawkins, *The Blind Watchmaker: Why the Evidence of Evolution Reveals a Universe Without Design* (New York: Norton, 1996).

13 Ibid., 296. As any architect knows, Dawkins could have retained the building metaphor by talking about specifications rather than recipes.

14 Ibid.

15 Kay, *Who Wrote the Book of Life? A History of the Genetic Code*, 2. She could have said 'data', or 'look up table' or 'index', as found at the back of a book.

16 Ibid.

17 Andrew Hodges, *Alan Turing: The Enigma of Intelligence* (London: Unwin Paperbacks, 1985).

18 In biology, however, the 'code' is the set of rules that translate DNA sequences to enzymes and then to characteristics and behaviours of the cell and the whole organism. In biology, the genome is not the code. The code is the hidden rules that translate DNA into something else.

19 Allan M. Turing, 'Computing Machinery and Intelligence', in *Computers and Thought*, ed. E. A. Feigenbaum and J. Feldman (Cambridge, MA: MIT Press), 11–35. For a helpful worked example of a Turing machine calculation (subtracting two numbers), see Mark Chu-Carroll (2012), 'Turing Machines: What they are, what they aren't'. *Good Math, Bad Math (A Scientopia blog)*, 24 June. Available online: http://goodmath.scientopia.org/2012/06/24/turing-machines-what-they-are-what-they-arent/ (accessed 11 June 2017).

20 Ehud Shapiro and Yaakov Benenson, 'Bringing DNA Computers to Life', *Scientific American* 294, no. (2006): 44–51.

21 Ibid.

22 Ibid.

23 Kay, *Who Wrote the Book of Life? A History of the Genetic Code.*

24 Haraway, 'A Cyborg Manifesto: Science, Technology, and Socialist Feminism in the Late Twentieth Century'; N. Katherine Hayles, *How We Became Posthuman: Virtual Bodies in Cybernetics, Literature, and Informatics* (Chicago, IL: University of Chicago Press, 1999).

25 Anton Markoš, *Readers of the Book of Life: Contextualizing Developmental Evolutionary Biology* (New York, NY: Oxford University Press, 2002).

26 Ibid.

27 Heidegger scholars will be familiar with the term *Dasein* to describe this entity, as explained by Heidegger, *Being and Time*. Also see Dreyfus, *Being-in-the-world: A Commentary on Heidegger's Being and Time, Division I.*

28 Also see Stephen J. Cowley, 'Bio-ecology and Language: A Necessary Unity', *Language Sciences* 41, no. (2014): 60–70.

29 Markoš, *Readers of the Book of Life: Contextualizing Developmental Evolutionary Biology.*

30 Ibid.

31 Ibid.

32 Ibid.

33 Ibid.

34 Markoš's recruitment of hermeneutics differs from what hermeneutical scholarship claims a reader does when they interpret a text. For a discussion of the relationship between hermeneutics and Peirce's concept of the interpretant, see Paul Ricoeur, 'What is a Text: Explanation and Interpretation', in *Myth-Symbolic Language and Philosophical Anthropology*, ed. D. M. Rasmussen (The Hague, Netherlands: Martinus Nijhoff), 135–50. Also see Kay, *Who Wrote the Book of Life? A History of the Genetic Code*, 36.

35 Dawkins, *The Blind Watchmaker: Why the Evidence of Evolution Reveals a Universe Without Design*, 158.

36 A gene is a fragment of DNA. We talk about a gene for green eyes, or blood type. Apparently, the DNA strand is the replicator. I have assumed the replicator is the machine that replicates the tweet/meme, and is therefore the human tweeter. Reading the literature, I think the text of the tweet is the replicator, like a fragment of DNA. The person, machine, book or website that stores and reproduces the meme is a vehicle (or interactor). Calling on Dawkin's definitions, Blackmore explains that 'Vehicles or interactors carry the replicators around inside them and protect them' (Susan Blackmore, *The Meme Machine* (Oxford, England: Oxford University Press, 1999), 5).

37 Ibid.

38 John James's website at www.johnjames.com.au (accessed 3 June 2017) includes a list of Gothic memes.

39 John James, *The Template-Makers of the Paris Basin* (Leura, Australia: West Grinstead, 1989), 4.

40 Paul-Alan Johnson, *The Theory of Architecture: Concepts, Themes and Practices* (New York: John Wiley, 1994).

41 Abel, *The Extended Self: Architecture, Memes and Minds*, 156. Also see further elaboration in Chris Abel, *Architecture and Identity: Responses to Cultural and Technological Change* (Abingdon, Oxon: Routledge, 2017), 98–109.

42 Abel, *The Extended Self: Architecture, Memes and Minds*, 162.

43 Richard Dawkins, *The Selfish Gene* (New York: Oxford University Press, 1976).

44 Haraway, *Simians, Cyborgs, and Women: The Reinvention of Nature*; Hayles, *How We Became Posthuman: Virtual Bodies in Cybernetics, Literature, and Informatics.*

45 Blackmore, *The Meme Machine*; Kim Zetter (2008), 'Humans are Just Machines for Propagating Memes (interview with Susan Blackmore)'. *Wired Magazine*, 29 February. Available online: https://www.wired.com/2008/02/ted-blackmore/?currentPage=all (accessed 11 June 2017). According to meme theory, memes combine and recombine, but there's nothing comparable to biological gendering and sexual reproduction as a way of ensuring

diversity in the gene pool. There are many other disanalogies of course identified in the meme literature. See Brandon Morrow (2016), 'Does Donald Trump Run His Own Twitter Account? 5 Fast Facts You Need to Know'. *Heavy*, 2 June. Available online: http://heavy.com/news/2016/06/does-donald-trump-run-his-twitter-account-page-operate-use-post-write-update-campaign-controversy-retweet-tweet-white-supremacist-bots-followers-first/ (accessed 11 June 2017).

46 Hans-Georg Gadamer, *Truth and Method* (New York: Seabury Press, 1975), 345.

47 Rachel Armstrong, *Vibrant Architecture: Matter as a Codesigner of Living Structures* (Warsaw: Gruyter Open, 2015).

48 Ibid., 41.

49 Ibid., 23.

50 The list includes ecological living technology, synthetic biology, protocell, natural computing, morphological computing, vibrant matter and inorganic sympathy.

51 Charissa N. Terranova and Meredith Tromble (eds), *The Routledge Companion to Biology in Art and Architecture* (London: Routledge, 2016).

52 The Latin *nātūra* (nature) is arguably less linguistically productive, offering fewer possibilities for hybridization, and *physicus* (natural science) fares little better.

Chapter 5

1 Immanuel Kant, *Observations on the Feeling of the Beautiful and the Sublime* (Berkeley, CA: University of California Press, 1960); Edmund Burke and James Boulton (ed.), *A Philosophical Enquiry into the Origin of our Ideas of the Sublime and Beautiful* (Notre Dame: University of Notre Dame Press, 1958). For a more recent treatment of landscape aesthetics, see Mitchell, 'Imperial Landscape'.

2 E. O. Wilson, *Biophilia* (Cambridge, MA: Harvard University Press, 1984).

3 René Magritte's painting of a pipe, with the caption 'This is not a pipe' is an obvious example of the play on signs evident in twentieth-century art. Also see John Berger, *Ways of Seeing* (London: Penguin, 1972).

4 Charles Jencks, *Iconic Building* (New York: Rizzoli, 2005) Iconoclasm involves the destruction of an icon, and Jencks makes references to the targeting of iconic structures in religious disputes, and recently by terrorists.

5 Jencks, *The Architecture of the Jumping Universe, A Polemic: How Complexity Science Is Changing Architecture and Culture.*

6 Perez-Gomez, *Attunement: Architectural Meaning After the Crisis of Modern Science*, 32.

7 Project Architect: Jürgen Mayer H., Andre Santer, Marta Ramírez Iglesias.

8 Gilles Deleuze and Félix Guattari, *A Thousand Plateaus: Capitalism and Schizophrenia* (London: Athlone Press, 1988).

9 Christopher Alexander, 'A City is not a Tree', in *Design After Modernism*, ed. J. Thackara (London: Thames and Hudson), 67–84.

10 Stephen R. Kellert, 'Dimensions, Elements, and Attributes of Biophilic Design', in *Biophilic Design: The Theory, Science and Practice of Bringing Buildings to Life*, ed. S. R. Kellert, J. Heerwagen and M. Mador (Hoboken, NJ: Wiley), 3–19, 10.

11 Ibid.

12 Ibid.

13 Thomas, *Technobiophilia: Nature and Cyberspace.*

14 Kent Bloomer, 'The Picture Window: The Problem of Viewing Nature Through Glass', in *Biophilic Design: The Theory, Science and Practice of Bringing Buildings to Life*, ed. S. R. Kellert, J. Heerwagen and M. Mador (Hoboken, NJ: Wiley), 253–62.

15 Ibid.

16 In fact, some architects skilfully shape and position windows specifically to point at (i.e. direct the occupants towards) objects in view: the mountaintop, the gazebo, the stone bridge, the meadow.

17 Kaplan, 'The Restorative Benefits of Nature: Toward an Integrative Framework'; R Kaplan and S Kaplan, *The Experience of Nature: A Psychological Perspective* (Cambridge, UK: Cambridge University Press, 1989).

18 Thomas, *Technobiophilia: Nature and Cyberspace,* loc921.

19 Peter H. Kahn, *Technological Nature: Adaptation and the Future of Human Life* (Cambridge, MA: MIT Press, 2011), xvi.

20 See Chapter 1 for this example of attention as identified by James, *The Principles of Psychology Volume I,* 395.

21 Peirce, *The Collected Papers of Charles Sanders Peirce. Vol. 2 Elements of Logic,* 52. This theme is developed further (without reference to Peirce) by the philosopher Jacques Derrida. See Jacques Derrida, *Of Grammatology* (Baltimore, MD: Johns Hopkins University Press, 1976). Apparently, Derrida studied Peirce's work but gives it only scant mention. See David E. Pettigrew, 'Peirce and Derrida: From Sign to Sign', in *Peirce's Doctrine of Signs: Theory, Applications, and Connections*, ed. V. M. Colapietro and T. M. Olshewsky (Berlin: Mouton de Gruyter), 365–78. We might think that the end point of such a referential chain is the meaning of the sign, but insofar as meaning resides anywhere it is for Derrida in the trace left by these chains of signification. See Jacques Derrida, *The Postcard: From Socrates to Freud and Beyond* (Chicago, IL: Chicago University Press, 1979). Nature provides a rich source of metaphors for how language operates. Trace, trail, track, wake, residue, and other natural remnants are evident in spider webs, the growth rings of a tree, layers of skin, carapaces, scales, streams, corals and other living and dead efflorescences, flows and secretions in nature.

22 Roland Barthes, *Mythologies* (London: Paladin, 1973).

23 Thirsten Klooster, *Smart Surfaces and Their Application in Architecture and Design* (Basel, Switzerland: Birkhäuser, 2009).

24 Ibid., 73.

25 Ibid.

26 Ludwig Bertalanffy, *General System Theory: Foundations, Development, Applications* (New York: G. Braziller, 1969).

27 Niklas Luhmann, *Social Systems* (Stanford, CA: Stanford University Press, 1995).

28 Sebeok, *Signs: An Introduction to Semiotics,* 34.

29 Klooster, *Smart Surfaces and Their Application in Architecture and Design,* 141.

30 Peter Bogh Andersen, 'Computer Semiotics', *Scandinavian Journal of Information Systems* 4, no. 1 (1992): 3–30, 18.

31 Ibid., 14.

32 Ibid.

33 J. Shaw, 'The Dis-Embodied Re-Embodied Body', *Kunstforum, Die Zukunft des Körpers I* 132, no. (1996); Andy Clark, *Being There: Putting Brain, Body and World Together Again* (Cambridge, MA: MIT Press, 1997); P. Dourish, *Where the Action Is: The Foundations of Embodied Interaction* (Cambridge, MA: MIT Press, 2001); Caroline A. Jones, *Sensorium: Embodied Experience, Technology, and Contemporary Art* (Cambridge, MA: MIT Press, 2006); Malcolm McCullough, *Ambient Commons: Attention in the Age of Embodied Information* (Cambridge, MA: MIT Press, 2013).

34 Herbert Simon, *The Sciences of the Artificial* (Cambridge, MA: MIT Press, 1969); Allen Newell and Herbert Simon, *Human Problem Solving* (Englewood Cliffs, NJ: Prentice-Hall, 1972).

35 Perez-Gomez, *Attunement: Architectural Meaning After the Crisis of Modern Science*, 32.

36 Lionel March (ed.), *The Architecture of Form* (London: Cambridge University Press, 1976).

37 For fuller accounts, see Robert Woodbury, *Elements of Parametric Design* (London: Routledge, 2010); Wassim Jabi, *Parametric Design for Architecture* (London: Laurence King, 2013).

38 Alan Redmond, et al., 'Exploring How Information Exchanges can be Enhanced Through Cloud BIM', *Automation in Construction* 24, no. (2012): 175–83.

39 William J. Mitchell, *Computer-Aided Architectural Design* (New York, NY: John Wiley and Sons, 1977); Antony D. Radford and Gary Stevens, *Computer Aided Design Made Easy: A Comprehensive Guide for Architects and Designers* (New York: Mcgraw-Hill, 1986); Yehuda E Kalay, *Architecture's New Media: Principles, Theories, and Methods of Computer-Aided Design* (London: MIT Press, 2004); Coyne, et al., *Knowledge-Based Design Systems.*

40 Philip Steadman, *Architectural Morphology: An Introduction to the Geometry of Building Plans* (London: Pion, 1983).

41 C. W. Churchman, 'Wicked Problems', *Management Science* 14, no. 4 (1967): B141–B2; Richard Buchanan, 'Wicked Problems in Design Thinking', in *The Idea of Design*, ed. V. Margolin and R. Buchanan (Cambridge, MA: MIT Press), 3–20; Richard Coyne, 'Wicked Problems Revisited', *Design Studies* 26, no. 1 (2005): 5–17.

42 See, for example, the design work of Archim Menges (http://www.achimmenges.net (accessed 3 June 2017)) and Greg Lynn, *Animate Form* (New York: Princeton Architectural Press, 1999); Greg Lynn, *Architecture for an Embryologic Housing* (Berlin: Birkhauser Verlag AG, 2002); Greg Lynn (ed.), *Folding in Architecture (Revised Edition)* (Chichester, England: Wiley-Academy, 2004), http://glform.com (accessed 5 June 2017).

43 Patrik Schumacher (2008), 'Parametricism as Style: Parametricist Manifesto'. *Writings*, Presentation at the Dark Side Club, 11th Architecture Biennale, Venice. Available online: http://www.patrikschumacher.com/Texts/Parametricism%20as%20Style.htm (accessed 11 June 2013); Patrik Schumacher, 'Parametricism: A New Global Style for Architecture and Urban Design', *Architectural Design* 79, no. 4 (2009): 14–23.

44 Schumacher, 'Parametricism: A New Global Style for Architecture and Urban Design'.

45 Ibid.

46 Ibid.

47 Mitchell, *Computer-Aided Architectural Design.*

48 Penrose, *The Emperor's New Mind: Concerning Computers, Minds, and the Laws of Physics*, 205. Parametricism also resonates with Stephen Hawking's controversial claims to seek a theory of everything (TOE). See Stephen Hawking, *A Brief History of Time: From the Big Bang to Black Holes* (New York: Bantam, 1988).

49 Cumincad, the online Cumulative Index of Computer-Aided Architectural Design (http://papers.cumincad.org (accessed 3 June 2017)), reveals over 700 books and articles on parametric design in architecture, dating back to the 1970s.

50 Marcos Cruz and Richard Beckett, 'Bioreceptive Design: A Novel Approach to Bio-digital Materiality', *Arq (Architectural Research Quarterley)* 20, no. 1 (2016): 51–64, 53. Also see Marcos Cruz, *The Inhabitable Flesh of Architecture* (Farnham, Surrey, England: Ashgate, 2013).

51 Cruz and Beckett, 'Bioreceptive Design: A Novel Approach to Bio-digital Materiality', 53.

52 Ibid.

53 Ibid., 62.

54 Markus Kayser, et al. (2013), 'Silk Pavillion'. *Mediated Matter*. Available online: http://matter.media.mit.edu/environments/details/silk-pavillion (accessed 4 June 2017).

55 Emma Flynn, '(Experimenting with) Living Architecture: A Practice Perspective', *Arq (Architectural Research Quarterley)* 20, no. 1 (2016): 21–8.

56 Ibid., 26.

57 Armstrong, *Vibrant Architecture: Matter as a Codesigner of Living Structures*, 55.

58 Ibid., 178.

59 Ibid., 179.

60 Crowe, *The Landscape of Power*.

61 Heidegger, *The Question Concerning Technology and Other Essays*, 152.

62 Don Ihde, *Postphenomenology: Essays in the Postmodern Context* (Evanston, IL: Northwestern University Press, 1995).

63 For further elaboration of the relationship between interpretation theory and pragmatism, see Endre Begby, 'Hermeneutics and Pragmatism', in *The Routledge Companion to Hermeneutics*, ed. J. Malpas and H.-H. Gander (Abingdon, Oxon: Routledge), 612–22.

Chapter 6

1 Jamie A. Davies, 'Machines for Living in: Connections and Contrasts between Designed Architecture and the Development of Living Forms', *Arq (Architectural Research Quarterley)* 20, no. 1 (2016): 45–50.

2 Ibid., 48.

3 Ibid.

4 John M Johansen (2011), 'Nanoarchitecture'. A Discourse Part I. Available online: http://johnmjohansen.com/Nanoarchitecture.html (accessed 11 June 2017).

5 Davies, 'Machines for Living in: Connections and Contrasts between Designed Architecture and the Development of Living Forms', 49.

6 Ibid.

7 https://soundcloud.com/artangel-2/master-rock (accessed 3 June 2017). See also a review by Claire Walsh (2015), 'The Hollow Mountain: Voice in the Performance of Maria Fusco's "Master Rock"'. *Map Magazine*. Available online: http://mapmagazine.co.uk/9890/hollow-mountain/ (accessed 11 June 2017).

8 Victor R. Baker, 'Geosemiosis', *Bulletin of the Geological Society of America* 111, no. 5 (1999): 633–45, 633.

9 Ibid., 638.

10 Ibid.

11 Ibid., 641.

12 Ibid.

13 For a summary of the countervailing view that there is a sharp division between life and non-life, see A. A. Sharov and T. Vehkavaara, 'Protosemiosis: Agency with Reduced Representation Capacity', *Biosemiotics* 8, no. 1 (2014): 103–23.

14 David Thomas Ansted, *The Great Stone Book of Nature* (London: Macmillan & Co., 1863), 46. For an account of the vocal attributes of glaciers, see Sverker Sölin, 'Do Glaciers Speak: The Political Aesthetics of vo/ice', in *Methodological Challenges in Nature-Culture and Environmental History Research*, ed. J. Thorpe, S. Rutherford and L. A. Sandberg (New York: Routledge), 13–30.

15 Ansted, *The Great Stone Book of Nature,* 46. Presumably movements of gasses, tectonic forces, radiation and impacts provide similar communicative functions for extraterrestrial semiotics.

16 Szerszynski, 'The End of the End of Nature: The Anthropocene and the Fate of the Human'.

17 Ibid., 166. *Erathems* and *eonothems* are deposits and stratified rocks laid down over particular time periods.

18 Ibid.

19 Ibid., 178.

20 Ibid. For a recent example of how the concept of the Anthropocene frames natural resource management issues, see Ray Ison, 'Governing in the Anthropocene: What Future Systems Thinking in Practice?', *Systems Research and Behavioural Science* 33, no. 5 (2016): 595–613.

21 See Timothy Clark, *Ecocriticism on the Edge: The Anthropocene as a Threshold Concept* (London: Bloomsbury, 2015) Jussi Parikka, *A Geology of Media* (Minneapolis, MN: University of Minnesota Press, 2015).

22 Simon L. Lewis and Mark A. Maslin, 'Defining the Anthropocene', *Nature* 519, no. (2015): 170–80, 171.

23 Szerszynski, 'The End of the End of Nature: The Anthropocene and the Fate of the Human'.

24 Ibid., 169.

25 Jan Zalasiewicz, et al., 'Are We Now Living in the Anthropocene?', *GSA (Geological Society of America) Today* 18, no. 2 (2008): 4–8.

26 Ibid., 4.

27 Ibid., 7.

28 Szerszynski, 'The End of the End of Nature: The Anthropocene and the Fate of the Human', 170

29 Ibid., 171.

30 Ibid., 180.

31 Ibid.

32 Thomas F Thornton and Yadvinder Malhi, 'The Trickster in the Anthropocene', *The Anthropocene Review* 3, no. 3 (2016): 201–4.

33 Ibid., 2. For an account of the 'digital anthropocene' that folds big data and the digital humanities into anthropocene discourse, see Charles Travis and Poul Holm, 'The Digital Environmental Humanities – What Is It and Why Do We Need It? The NorFish Project and SmartCity Lifeworlds', in *The Digital Arts and Humanities: Neography, Social Media and Big Data Integrations and Applications*, ed. C. Travis and A. von Lünen (AG Switzerland: Springer), 187–204.

34 Thornton and Malhi, 'The Trickster in the Anthropocene', 2.

35 Ibid., 7.

36 Szerszynski takes the example of the steersman from Michel Serres, who in turn borrows the term from cybernetics. See Michel Serres and Felicia McCarren, 'The Natural Contract', *Critical Inquiry* 19, no. 1 (1992): 1–21 and Norbert Weiner, *The Human Use of Human Beings* (Cambridge, MA: Da Capo Press, 1950).

37 Serres and McCarren, 'The Natural Contract', 14

38 Ibid.

39 Lesley Head, *Hope and Grief in the Anthropocene: Re-conceptualising Human–nature Relations* (Abingdon, Oxon: Routledge, 2016), 167.

40 Ibid., 169.

41 Ibid., 168.

42 Ibid., 170.

43 Ibid.

44 Ibid., 171.

45 Ibid.

46 Ibid., 173.

47 From a different standpoint, Serres discusses the human relationship with nature in terms of 'a natural contract of symbiosis and reciprocity' (Serres and McCarren, 'The Natural Contract', 11). He adds: 'The Earth speaks to us in terms of force, bonds, and interactions, and that suffices to make a contract' (ibid., 12).

48 On emotional labourers and their problems, see Arlie Russell Hochschild, *The Managed Heart: Commercialization of Human Feeling* (Berkeley, CA: University of California Press, 2012).

49 Head, *Hope and Grief in the Anthropocene: Re-conceptualising Human–nature Relations,* 167.

50 Consider the appointment of a climate change sceptic to the Head of the US Environment Protection Agency in 2017. Also see McKibben, 'The End of Nature'.

51 Head, *Hope and Grief in the Anthropocene: Re-conceptualising Human–nature Relations,* 168.

52 Ibid.

53 Ibid., 167.

54 Ibid. Artist and environmentalist Maya Lin memorializes loss in the Anthropocene. See http://whatismissing.net/ (accessed 26 February 2017). Some term the psychological distress that comes from environmental change 'solastalgia'. See Glenn Albrecht, 'Solastalgia: A New Concept in Health and Identity', *PAN: Philosophy, Activism, Nature* 3, no. (2005): 44–59.

55 Head, *Hope and Grief in the Anthropocene: Re-conceptualising Human–nature Relations,* 10.

56 A BBC nature documentary series called *Spy in the Wild* (John Downer Productions, 2017) shows monkeys apparently grieving over the 'death' of a synthetic monkey automaton, and herds of giraffes coming to the carcass of a dead giraffe as if to pay respects.

57 For a review article on extinction, emotion and stress in animals, see Richard Smyth, 'Happy Planet?', *BBC Wildlife* 35, no. 6 (2017): 31–4.

58 Head, *Hope and Grief in the Anthropocene: Re-conceptualising Human–nature Relations,* 25.

59 Rom. 8.22.

60 Coyne, *Mood and Mobility: Navigating the Emotional Spaces of Digital Social Networks.*

61 Head, *Hope and Grief in the Anthropocene: Re-conceptualising Human–nature Relations,* 22.

62 Ibid.

63 Jonathan Flatley, *Affective Mapping: Melancholia and the Politics of Modernism* (Cambridge, MA: Harvard University Press, 2008).

64 In her influential book *On Death and Dying* Elizabeth Kübler-Ross outlines the five stages of transition by which someone deals with loss. See Elizabeth Kübler-Ross, *On Death and Dying: What the Dying Have to Teach Doctors, Nurses, Clergy and Their Own Families* (New York: Scribner, 2014).

65 Head, *Hope and Grief in the Anthropocene: Re-conceptualising Human–nature Relations,* 41.

66 Ibid.

67 Ibid.

68 Michelle Bastian, 'Fatally Confused: Telling the Time in the Midst of Ecological Crises', *Environmental Philosophy* 9, no. 1 (2012): 23–48.

69 Serres and McCarren, 'The Natural Contract', 6.

70 Ibid.

71 Ibid., 6-7.

72 Sigmund Freud, 'Mourning and Melancholia', in *The Standard Edition of the Complete Psychological Works of Sigmund Freud, Volume XIV (1914-1916): On the History of the Psycho-Analytic Movement, Papers on Metapsychology and Other Works* (London: Hogarth Press), 237–58.

73 Coyne, *Mood and Mobility: Navigating the Emotional Spaces of Digital Social Networks*; Richard Coyne, 'Melancholy Urbanism: Distant Horizons and the Presentation of Place', in *Cinematic Urban Geographies*, ed. F. Penz and R. Koeck (London: Palgrave Macmillan).

74 According to Walter Benjamin the horizon is the main motif in Albrecht Durer's depiction of the angel Melancholia. See Walter Benjamin, *The Origin of German Tragic Drama* (London: Verso, 2003).

75 For a similar strategy, see Michelle Bastian and Thom van Dooren, 'The New Immortals: Immortality and Infinitude in the Anthropocene', *Journal of Environmental Philosophy* 14, no. 1 (2017): 1–9. Szerszynski also wants to 'desecularize' the Anthropocene. See Bronislaw Szerszynski, 'Gods of the Anthropocene: Geo-spiritual Formations in the Earth's New Epoch', *Theory, Culture and Society* 34, no. 2–3 (2017): 253–75.

76 Architects of buildings made of 'earth bags' and straw bales claim a reconnection with the earth, for example, the work of Paulina Wojciechowska (http://earthhandsandhouses.org (accessed 3 April 2017)). See Marcin Kolakowski, 'Modernism or Tradition in Low-

technology? A Humanistic Perspective on the Architecture of Paulina Wojciechowska', *ACEE Architecture Civil Engineering* no. 1 (2016): 21–34.

77 Baker, 'Geosemiosis', 640

78 David Abram, *Becoming Animal: An Earthly Cosmology* (New York: Vintage, 2011), 63.

79 Ibid.

80 Ibid.

81 Ibid.

82 Environmental scientists speak of returning greenhouse gases (carbon compounds) from the atmosphere back to the ground – earth as 'carbon sink'. See United Nations / Framework Convention on Climate Change, *Adoption of the Paris Agreement, 21st Conference of the Parties* (Paris: United Nations, 2015).

83 John's Gospel delivers nascent support for the originary character of a semiotics of nature, in this case, *logos*: 'In the beginning was the word' (Jn 1.1).

84 Gen. 2.7.

85 Plato, Statesman.

86 Mary Shelley, *Frankenstein* (Basingstoke: Macmillan, 1992). Cyborg artist Stelarc's art is about body parts, body fluids and prosthetics. I discussed autochthony further in the context of Freud and Lacan in Coyne, *Technoromanticism: Digital Narrative, Holism, and the Romance of the Real*. On magma heroes and monsters emerging from the ground in contemporary comics, see Nigel Clark, 'Pyropolitics for a Planet of Fire', in *Territory Beyond Terra*, ed. K. Peters, P. Steinberg and E. Stratford (London: Rowman and Littlefield), to appear.

87 Thomas J. Morrissey and Richard Wunderlich, 'Death and Rebirth in Pinocchio', *Children's Literature* 11, no. (1983): 64–75.

88 Plato's account of autochthony includes a return to the earth in old age: 'It's in keeping with the idea of old people turning into children that people would reform in the earth where they were lying after their death and would come back to life from there, in conformity with the reversal undergone by all natural cycles'. Plato, *Statesman* (Cambridge, England: Cambridge University Press, 1995), 24.

89 Lévi-Strauss explained the age-old Oedipus myth in terms of the conflict between life born of the earth (autochthony) and that emerging from sexual reproduction.

90 Claude Lévi-Strauss, *Structural Anthropology 1* (New York: Basic Books, 1963), 216.

91 Ibid.

92 The goddess, Gaia, Mother Earth, gave birth to the hills and the sea, born without a father. See Hesiod, *Theogony and Works and Days* (Oxford: Oxford University Press, 1988), 129–32. For a further account relating Greek gods to the Anthropocene, see Bastian and van Dooren, 'The New Immortals: Immortality and Infinitude in the Anthropocene'.

93 Thornton and Malhi, 'The Trickster in the Anthropocene', 2.

94 In his documentation of the sagas, Snorri Sturlson (1179–1241) said: 'Loki is pleasing, even beautiful to look at, but his nature is evil and he is undependable.' Snorri Sturluson, *The Prose Edda: Norse Mythology* (London: Penguin, 2006), 918.

95 Stefanie von Schnurbein, 'The Function of Loki in Snorri Sturluson's "Edda"', *History of Religions* 40, no. 2 (2000): 109–24.

96 Ibid., 115.

97 Digital interaction designers reference the earth. See Malcolm McCullough, *Digital Ground: Architecture, Pervasive Computing, and Environmental Knowing* (Cambridge, MA: MIT Press, 2004).

98 'Lulz' is what it sounds like when you try to pronounce LOL (laugh out loud) in the plural, and 'Sec' more obviously abbreviates 'security'.

99 Nate Anderson (2011), 'LulzSec Manifesto: "We Screw Each other Over for a Jolt of Satisfaction"'. *Ars Technica*, 17 June. Available online: https://arstechnica.com/tech-policy/2011/06/lulzsec-heres-why-we-hack-you-bitches/ (accessed 3 June 2017).

100 Ibid.

101 Lewis Hyde, *Trickster Makes This World: Mischief, Myth and Art* (New York: North Point Press, 1998). I examined the trickster function in relation to digital commerce in Coyne, *Cornucopia Limited: Design and Dissent on the Internet.*

102 Jason Byrne and Natalie Osborne (2016), 'Urban Hacktivism: Getting Creative About Involving Citizens in City Planning'. *The Conversation (blog site)*, 5 July. Available online: https://theconversation.com/urban-hacktivism-getting-creative-about-involving-citizens-in-city-planning-62277 (accessed 3 June 2017). For a recent summary of urban activist practices, see Lucy (ed.) Bullivant, *4D Hyper-Local: A Cultural Tool Kit for the Open Source City* (Chichester, UK: Wiley, 2017).

103 Michael Hardman (2014), 'Look Out Behind the Bus Stop, here come Guerrilla Gardeners Digging up an Urban Revolution'. *The Conversation (blog site)*, 16 June. Available online: https://theconversation.com/look-out-behind-the-bus-stop-here-come-guerrilla-gardeners-digging-up-an-urban-revolution-29225 (accessed 3 June 2017). On community gardening as radical urban practice, see Tahl Kaminer, *The Efficacy of Architecture: Political Contestation and Agency* (London: Routledge, 2017), 108–20.

104 For a summary of staid environmental digital tactics, see A. D. Thaler, et al., 'Digital Environmentalism: Tools and Strategies for the Evolving Online Ecosystem', in *Environmental Leadership: A Reference Handbook*, ed. E. Gallagher (Los Angeles, CA: Sage), 364–72.

105 For Jung, 'it is always the father-figure from whom the decisive convictions, prohibitions, and wise counsels emanate'. Carl G. Jung, *Four Archetypes: Mother, Rebirth, Spirit, Trickster* (London: Ark, 1986), 92.

106 Thornton and Malhi, 'The Trickster in the Anthropocene', 3.

107 Ibid.

108 Ibid.

109 Ibid.

110 Ibid., 4.

111 Coyne, *Cornucopia Limited: Design and Dissent on the Internet.*

Chapter 7

1 John Wills, 'Digital Dinosaurs and Artificial Life: Exploring the Culture of Nature in Computer and Video Games', *Cultural Values* 6, no. 4 (2002): 395–417, 411.

2 Ibid.

3 Ibid.

4 Ibid., 397.

5 Ibid., 398.

6 Thomas Hobbes, *Leviathan* (Cambridge, England: Cambridge University Press, 1991), 88.

7 Ibid., 89.

8 Researchers have also reviewed certain video games for their therapeutic value in improving mental health. For a critical study, see M Brown, et al., 'Gamification and Adherence to Web-based Mental Health Interventions: A Systematic Review', *JMIR Mental Health* 3, no. 3 (2016):

9 Hobbes, *Leviathan*, 399.

10 Ibid.

11 Emerson, *Essays by Ralph Waldo Emerson*, 12.

12 Hobbes, *Leviathan*, 400.

13 Ibid., 401.

14 Ibid., 403.

15 Hayles, 'Narratives of Artificial Life'.

16 Wills, 'Digital Dinosaurs and Artificial Life: Exploring the Culture of Nature in Computer and Video Games', 409.

17 Ibid., 410.

18 Mary Flanagan, *Critical Play: Radical Game Design* (Cambridge, MA: MIT Press, 2009), 2.

19 Hanna Drummond, *The Application of Cognitive Behavioural Therapy in Interactive Media* (Edinburgh, UK: Unpublished Research Report, The University of Edinburgh, 2016) See the game website at http://thatgamecompany.com/games/journey/ (accessed 3 June 2017).

20 One of the challenges of such analysis is to identify how long such effects last.

21 Wills, 'Digital Dinosaurs and Artificial Life: Exploring the Culture of Nature in Computer and Video Games', 412.

22 Ibid.

23 Ibid.

24 Hannah Gould (2016), 'If Pokémon Go Feels Like a Religion, that's Because it Kind of is'. *Guardian*, 3 June 2017. Available online: https://www.theguardian.com/technology/2016/jul/12/pokemon-go-addictive-game-shares-much-with-religious-devotion (accessed 12 July 2016).

25 Tillmann Prüfer and Sebastian Mondial, '"Ingress": When Google Plays Games in a Concentration Camp'. *Zeit Magazin*, 1 July (2015). Available online: http://www.zeit.de/zeit-magazin/leben/2015-07/ingress-smartphone-game-google-niantic-labs-nazis-concentration-camp (accessed 3 June 2017).

26 Julian Stallabras, *Gargantua: Manufactured Mass Culture* (London: Verso, 1996).

27 Jean Baudrillard, *Simulacra and Simulation* (Ann Arbor: University of Michigan press, 1994).

28 As identified by the philosopher Henri Lefebvre. See Henri Lefebvre, *The Production of Space* (Oxford, UK: Blackwell, 1991).

29 Lefebvre states: 'The dominant tendency fragments space and cuts it into pieces. It enumerates the things, the various objects that space contains.' (ibid., 89.).

30 Comment posted by Dynasty2021 12 Jul 2016 8:56 to Gould, 'If Pokémon Go Feels Like a Religion, that's Because it Kind of is'.

31 Flanagan, *Critical Play: Radical Game Design.*

32 Jeffrey L. Kidder, 'Parkour, the Affective Appropriation of Urban Spaces, and the Real/Virtual Dialectic', *City and Community* 11, no. 3 (2012): 229–53, 244.

33 Ibid., 246.

34 Ibid., 247.

35 Ibid.

36 Ibid., 248.

37 Michael Atkinson, 'Parkour, Anarcho-environmentalism, and Poiesis', *Journal of Sport and Social Issues* 33, no. 2 (2009): 169–94.

38 Ibid., 4.

39 Ibid.

40 Thoreau, *Walden: Or, Life in the Woods.*

41 Atkinson, 'Parkour, Anarcho-environmentalism, and Poiesis', 11.

42 Ibid.

43 Ibid., 16.

44 Ibid., 17.

45 See the project website at https://www.gardenbridge.london/ (accessed 13 March 2017).

46 Marshall McLuhan wrote a great deal about urban tribes in the electronic age, for example, 'the return to Nature and the return to the tribe are under electric conditions, fatally simple'. Marshall McLuhan, *Understanding Media, The Extensions of Man* (Cambridge, MA: MIT Press, 1994), 155.

47 For Peirce, a metaphor is a kind of iconic sign. As identified by R. Lance Factor, Peirce thought that most of language could be accounted for sufficiently by the idea of metaphor. R. Lance Factor, 'Peirce's Definition of Metaphor and its Consequences', in *Peirce's Doctine of Signs: Theory, Application, and Connections*, ed. V. M. Colapietro and T. M. Olshewsky (Berlin: Mouton de Gruyter), 229–35, 229.

48 Jonathan Conlin, *The Pleasure Garden, from Vauxhall to Coney Island* (Philadelphia, PA: University of Pennsylvania Press, 2013), 1.

49 Ibid.

50 Tia DeNora, *Music Asylums: Wellbeing Through Music in Everyday Life* (Farnham, England: Ashgate, 2013), 42.

51 Ibid., 42–3.

52 Ibid., 43.

53 Roger Caillois, *Man, Play, and Games* (New York, NY: The Free Press of Glencoe, 1961).

54 Ibid., 14.

55 Alfred Lord Tennyson in the poem *In Memoriam A.H.H.* (1850).

56 Charles Darwin, *The Origin of Species* (Oxford: Oxford University Press, 1996).

57 Wendy Pullan, 'Agon in Urban Conflict: Some Possibilities', in *Phenomenologies of the City: Studies in the History and Philosophy of Architecture*, ed. H. Steiner and M. Sternberg (Farnham, England: Ashgate), 213–24.

58 Johan Huizinga, *Homo Ludens: A Study of the Play Element in Culture* (Boston: Beacon Press, 1955).

59 Ibid., 222.

60 Hans-Georg Gadamer, *Truth and Method* (New York: Continuum, 2004).

61 See Snodgrass and Coyne, *Interpretation in Architecture: Design as a Way of Thinking.*

62 Huizinga, *Homo Ludens: A Study of the Play Element in Culture.*

63 Gadamer, *Truth and Method,* 104.

64 Ibid.

65 Ibid., 105. The italics are in the original.

66 Ibid., 103.

67 Coyne, *Cornucopia Limited: Design and Dissent on the Internet.*

68 Saussure, *Course in General Linguistics.*

69 Douglas Rand Anderson, *Creativity and the Philosophy of C. S. Peirce* (Dordrecht, The Netherlands: Springer, 1987).

70 Peirce, *The Collected Papers of Charles Sanders Peirce. Vol. 5 Pragmatism and Pragmaticism*; Anderson, *Creativity and the Philosophy of C. S. Peirce.*

71 Ibid., 475.

72 Ibid.

73 Jacques Derrida, 'Différance', in *Margins of Philosophy* (Chicago: University of Chicago press), 3–27; Coyne, *Derrida for Architects.*

74 Jacques Derrida, *Aporias* (Stanford, CA: Stanford University Press, 1993).

75 As it happens, Derrida makes only scant reference to Peirce in his writing. Jacques Derrida, 'The Principle of Reason: The University in the Eyes of its Pupils', *Diacritics* 13 (1983): 3–20; Christopher Norris, *Derrida* (Cambridge, MA: Harvard University Press, 1987).

76 Douglas R. Anderson, *Creativity and the Philosophy of C. S. Peirce (Abduction, Evolution, God)* (Ann Arbor, MI: PhD Thesis, Pennsylvania State University, 1984).

77 Peirce, *The Collected Papers of Charles Sanders Peirce. Vol. 5 Pragmatism and Pragmaticism*; March (ed.), *The Architecture of Form.*

78 Alan F. Chalmers, *What is this Thing Called Science?* (Indianapolis, IN: Hackett Pub Co, 1999).

79 See Churchman, 'Wicked Problems'; Horst Rittel and Melvin Webber, 'Dilemmas in a General Theory of Planning', *Policy Sciences* 4 (1973): 155–69; Buchanan, 'Wicked Problems in Design Thinking'; Coyne, 'Wicked Problems Revisited'. For an account of social media inspired democracy movements and wicked problems, see Evgeny Morozov, *The Net Delusion: How Not to Liberate the World* (London: Allen Lane, 2011).

80 Umberto Eco, *Semiotics and the Philosophy of Language* (London: Macmillan, 1984); Umberto Eco, *The Name of the Rose* (London: Vintage, 1980).

Chapter 8

1 Randy Malamud, 'Introduction: Famous Animals in Modern Culture', in *A Cultural History of Animals in the Modern Age*, ed. R. Malamud (Oxford: Berg), 1–26, 2. To this list we could add the way animal populations are introduced, culled and managed. For a brief account of deer populations in Scotland, see Patricia Macdonald, 'Change in Glen Feshie:

Environmental Change in a Dynamic Cairngorms Landscape', *The Nature of Scotland*, no. 24 (2016): 12–7.

2 Malamud, 'Introduction: Famous Animals in Modern Culture', 12.

3 See the Infinite Cat Project from 2006. http://www.infinitecat.com/.

4 'Dolphin Dies Being Passed around for Selfies' *Sky News*, 18 February (2016). Available online: http://news.sky.com/story/dolphin-dies-being-passed-around-for-selfies-10171630 (accessed 3 June 2017).

5 Harry Brenton, et al., 'The Uncanny Valley: Does it Exist?', in *Proceedings of Conference of Human Computer Interaction, Workshop on Human animated Character Interaction.*

6 Note the etymological link between *satire* and *satyr*, a creature that is part human and part animal.

7 Paul Shepard, *The Others: How Animals Made Us Human* (Washington, DC: Island Press, 1996).

8 Shepard, *The Others: How Animals Made Us Human*, 81–2.

9 An interesting article by cultural theorist Boria Sax confirms as much (Boria Sax, 'The Cosmic Spider and Her Worldwide Web: Sacred and Symbolic Animals in the Era of Change', in *A Cultural history of Animals in the Modern Age*, ed R. Malamud [Oxford: Berg], 27–48). As with our pets, we do not necessarily hunt our technologies, but we certainly play with them, or they play with and tease us, as if to make us look incompetent when we let them.

10 Shepard, *The Others: How Animals Made Us Human*, 46.

11 Ibid.

12 http://www.spore.com/ (accessed 3 June 2017).

13 Shepard, *The Others: How Animals Made Us Human*, 61.

14 The ancient Greek philosopher Diogenese the Cynic was an inhabitant of the streets and consorted with dogs. Canines return to their vomit, and copulate and defecate in public. See Farrand Sayre, *Diogenes of Sinope* (Baltimore: J. H. Furst Company, 1938); Donald Dudley, *A History of Cynicism: From Diogenes to the 6th Century AD* (Hildesheim, Germany: Georg Olms Verlagsbuchhandlung, 1967); Coyne, *Cornucopia Limited: Design and Dissent on the Internet*.

15 Shepard, *The Others: How Animals Made Us Human*, 60.

16 Ibid.

17 Ibid.

18 Ibid., 68.

19 The philosopher of biopolitics Giorgio Agamben identifies the religious idea of the resurrection of the body as a denial of our animal natures. I would say that we are dealing here with metaphors that support all kinds of impossibilities, but the point is well made. See Giorgio Agamben, *The Open: Man and Animal* (Stanford, CA: Stanford University Press, 2004).

20 Mark Kermode, 'Zootropolis Review: Disney's Animated Odd Couple has a Perfect Chemistry'. *Guardian*, 27 March (2016). Available online: https://www.theguardian.com/film/2016/mar/27/zootropolis-review-disney-animated-odd-couple-mark-kermode (accessed 3 June 2017).

21 Pollio Vitruvius, *Vitruvius: The Ten Books on Architecture* (New York: Dover Publications, 1960).

22 Commonly known as a 'greater road runner', as depicted in the famous Warner Brothers Roadrunner cartoons.

23 Hayles, *How We Became Posthuman: Virtual Bodies in Cybernetics, Literature, and Informatics* The position of the philosopher Giorgio Agamben has often been described as Posthumanist.

24 At least in his book *The Open: Man and Animals*, Agamben does not work through the difference this post-human orientation makes to our actions in the world. Nor does he relate this position to other contemporary thinking outside of the orbit of philosophy.

25 Samuel Gibbs, 'Apple Co-Founder Steve Wozniak Says Humans will be Robots' Pets'. *Guardian*, 25 June (2015). Available online: https://www.theguardian.com/technology/2015/jun/25/apple-co-founder-steve-wozniak-says-humans-will-be-robots-pets (accessed 3 June 2017).

26 Turkle, *Alone Together: Why We Expect More from Technology and Less from Each other.*

27 Shepard, *The Others: How Animals Made Us Human*, 143.

28 Ibid.

29 Ibid., 144.

30 Ibid.

31 John Berger, 'Why Zoos Disappoint', *New Society* 40, no. (1977): 122–3.

32 Shepard, *The Others: How Animals Made Us Human*, 145.

33 Ibid., 130132.

34 English literature and art theorist W. J. T. Mitchell provides an interesting observation about animals and vision. They 'see what we see', but are they convinced that a painting or photograph of a scene or of another animal is the same as what we see? W. J. T. Mitchell, *Picture Theory: Essays on Verbal and Visual Representation* (Chicago: University of Chicago Press, 1995), 334.

35 Shepard, *The Others: How Animals Made Us Human*, 131.

36 Ibid.

37 Ibid., 141.

38 Douglas, *Purity and Danger: An Analysis of Concepts of Pollution and Taboo.*

39 Max Black, *Models and Metaphors: Studies in Language and Philosophy* (Ithaca: Cornell University Press, 1962); Jacques Derrida, 'White Mythology: Metaphor in the Text of Philosophy', *New Literary History* 61 (1974): 5–74; Donald Schön, 'Generative Metaphor: A Perspective on Problem-Setting in Social Policy', in *Metaphor and Thought*, ed. A. Ortony (Cambridge: Cambridge University Press), 254–83; Richard Coyne, et al., 'Metaphors in the Design Studio', *JAE (Journal of Architectural Education)* 48, no. 2 (1994): 113–25.

40 Sianne Ngai, *Our Aesthetic Categories: Zany, Cute, Interesting* (Cambridge, MA: Harvard University Press, 2012), 87.

41 Ngai, *Our Aesthetic Categories*, 135.

42 Ibid., 64.

43 Jacques Derrida and David Wills, 'The Animal that Therefore I Am (More to follow)', *Critical Inquiry* 28, no. 2 (2002): 369–418.

44 Haraway, *Simians, Cyborgs, and Women: The Reinvention of Nature.*

45 Derrida also responds to the modernist Jeremy Bentham's question of whether animals can suffer. See Derrida and Wills, 'The Animal that Therefore I Am (More to follow)', 396. Heidegger puts animals beneath human beings. Agamben disagrees with Heidegger. For Agamben the mystery of the human–animal relationship resides in the 'practical and political mystery of separation'. Agamben, *The Open: Man and Animal,* loc168.

46 On the way groups, tribes and teams self-identify with whatever animals are to hand, see Lévi-Strauss, *Structural Anthropology 1.*

Chapter 9

1 I explored tuning at length in Coyne, *The Tuning of Place: Sociable Spaces and Pervasive Digital Media.* On attunement in architecture also see Perez-Gomez, *Attunement: Architectural Meaning After the Crisis of Modern Science.*

2 Houser, 'Introduction', xxxii–xxxiii

3 Ibid.

4 Perez-Gomez, *Attunement: Architectural Meaning After the Crisis of Modern Science,* 1.

5 Ibid., 90. For similar arguments about harmony in the urban environment, see Jonathan F. P. Rose, *The Well-Tempered City: What Modern Science, Ancient Civilizations, and Human Nature Teach Us About the Future of Urban Life* (London: HarperCollins, 2016).

6 Guignon, *The Good Life,* xiii.

7 John (ed.) Gaskin, *The Epicurean Philosophers* (London: Everyman, 1995).

8 Perez-Gomez, *Attunement: Architectural Meaning After the Crisis of Modern Science,* 75.

9 Houser, 'Introduction', xxxii.

10 Aurelius, *Meditations*; William L. Davidson, *The Stoic Creed* (New York: Arno Press, 1979); John Sellars, 'The Point of View of the Cosmos: Deleuze, Romanticism, Stoicism', *Pli (The Warwick Journal of Philosophy)* 8 (1999): 1–24; P. Clarke, 'Adam Smith, Stoicism and religion in the 18th century', *History of the Human Sciences* 13, no. 4 (2000): 49–72. I discuss Stoicism in the context of digital media in Coyne, *Cornucopia Limited: Design and Dissent on the Internet.*

11 Franklin Miller and Luana Colloca, 'Semiotics and the Placebo Effect', *Perspectives in Biology and Medicine* 53, no. 4 (2010): 509–16.

12 Jon Anderson, 'Transient Convergence and Relational Sensibility: Beyond the Modern Constitution of Nature', *Emotion, Space and Society* 2 (2009): 120–7.

13 Guignon, *The Good Life,* 2.

14 Ibid., 1.

15 Turkle, *Alone Together: Why We Expect More from Technology and Less from Each other,* The New State of the Self: Tethered and Marked Absent.

16 Perez-Gomez, *Attunement: Architectural Meaning After the Crisis of Modern Science,* 18.

17 Ibid.

18 Ibid.

19 'Pavement Lights Guide "Smartphone Zombies"', *BBC News,* 16 February (2017). Available online: http://www.bbc.co.uk/news/technology-38992653 (accessed 3 June 2017).

20 John E. Dunn, 'Smartphones Stress Users with "phantom" Text Messages'. *Techworld*, 12 January (2012). Available online: http://www.techworld.com/news/personal-tech/smartphones-stress-users-with-phantom-text-messages-3329685/ (accessed 3 June 2017).

21 'Smartphones can Increase Stress Levels, Study Says', *Huffpost*, 12 January (2012). Available online: http://www.huffingtonpost.com/2012/01/12/smartphones-cause-stress_n_1202924.html (accessed 3 June 2017).

22 Ofcom, *The Communications Market Report* (London: Ofcom, 2016), 4.

23 Ibid.

24 Ibid., 41.

25 As a semiotic play, the meaning of 'detox' has drifted over the years. It refers to 'toxins', which are poisons. 'To detox' has come to mean getting out of any condition of dependency or addiction. It suggests that balance is being restored.

26 T. Hartig, et al., 'Tracking Restoration in Natural and Urban Field Settings', *Journal of Environmental Psychology* 23 (2003): 109–23; Jennifer Roe and Peter Aspinall, 'The Restorative Benefits of Walking in Urban and Rural Settings in Adults with Good and Poor Mental Health', *Health and Place* 17 (2011): 103–13; R. S. Ulrich, et al., 'Stress Recovery During Exposure to Natural and Urban Environments', *Journal of Environmental Psychology* 11 (1991): 201–30.

27 Stephen Kaplan, 'The Restorative Benefits of Nature: Toward an Integrative Framework', ibid., 15 (1995): 169–82; Kaplan and Kaplan, *The Experience of Nature: A Psychological perspective.*

28 For reviews of the relationship between health and outdoor exercise, see T. Sugiyama and C. Ward Thompson, 'Older People's Health, Outdoor Activity and Supportiveness of Neighbourhood Environments', *Landscape and Urban Planning* 83 (2007): 168–75; Catharine Ward Thompson, et al., *Innovative Approaches for Researching Landscape and Health: Open Space: People Space 2* (Abingdon, England: Routledge, 2010); Catharine Ward Thompson, et al., 'More Green Space is Linked to Less Stress in Deprived Communities: Evidence from Salivary Cortisol Patterns', *Landscape and Urban Planning* 105 (2012): 221–9; Valerie F. Gladwell, et al., 'The Great Outdoors: How a Green Exercise Environment can Benefit All', *Extreme Physiology & Medicine* 2, no. 3 (2013): 1–7. For an evidence-based study into attitudes to walking, see Colin Pooley, et al., *Understanding Walking and Cycling: Summary of Key Findings and Recommendations* (Lancaster, England: Lancaster University, 2011).

29 Forestry Commision, *Woods for Health Strategy* (Edinburgh: Forestry Commission Scotland, 2009).

30 Timothy Beatley and Peter Newman, 'Biophilic Cities are Sustainable, Resilient Cities', *Sustainability* 5 (2013): 3328–45, 3335.

31 Beatley and Newman, 'Biophilic Cities are Sustainable, Resilient Cities', 3337.

32 Ibid., 3338.

33 Ibid.

34 Ibid., 3340.

35 Ibid., 3342.

36 Ibid., 3337.

37 On outdoor health apps, see T. Harries, et al., 'Walking in the Wild: Using an Always-on Smartphone Application to Increase Physical Activity', in *Lecture Notes in Computer Science, vol. 8120*, ed. P. Kotzé, et al. (Berlin: Springer), 19–36; Oresti Banos, et al. 'mHealthDroid: A Novel Framework for Agile Dev. of Mobile Health Apps', in *Lecture Notes in Computer*

Science, Vol 8868, ed. L. Pecchia, L. L. Chen, C. Nugent and J. Bravo, (Cham: Springer), 91–8; Tim Harries, et al., 'Effectiveness of a Smartphone App in Increasing Physical Activity Amongst Male Adults: A Randomised Controlled Trial', *BMC Public Health* 16, no. 925 (2016): 1–10.

38 Amid the myriad of photographic presentations of natural landscapes I will here refer to the aerial photography of altered wild landscapes by colleagues Patricia Macdonald and Angus Macdonald, 'Rephotography in the Scottish Highlands: Cairngorms & Morvern'. *Aerographica*, 7 June (2017). Available online: http://aerographica.org/re-photography-in-the-cairngorms-and-morvern-scottish-highlands/ (accessed 11 June 2017).

39 For a popular (and gendered) account of the human propensity to seek refuge, see John Gray, *Men are from Mars, Women are from Venus* (London: Harper Collins, 1992).

40 Thoreau, *Walden: Or, Life in the Woods.*

41 Adam Sharr, *Heidegger's Hut* (Cambridge, MA: MIT Press, 2006).

42 DeNora, *Music Asylums: Wellbeing Through Music in Everyday Life.*

43 Ibid., 56.

44 Erving Goffman, *Asylums: On the Social Situation of Mental Health Patients and Other Inmates* (New York: Anchor Books, 1962).

45 DeNora, *Music Asylums: Wellbeing Through Music in Everyday Life*, 63.

46 Inger E. Burnett-Zeigler, 'How Donald Trump Affects Therapy Patients'. *Time*, 28 November (2016). Available online: http://time.com/4583628/donald-trump-therapy-patients/ (accessed 3 June 2017).

47 Also see Ayesha Hazarika, 'Labour's Comedy Therapy, the Brexit Satire Boom and the Return of the Pink Bus', *New Statesman*, 26 August to 1 September (2016): 19.

48 Shaun Gallagher, *Hermeneutics and Education* (Albany, NY: SUNY Press, 1992).

49 Gilles Deleuze and Félix Guattari, '1837: Of the Refrain', in *A Thousand Plateaus: Capitalism and Schizophrenia* (London: Athlone Press), 310–50.

50 Ibid., ?.

51 Coyne, *Mood and Mobility: Navigating the Emotional Spaces of Digital Social Networks.*

52 DeNora, *Music Asylums: Wellbeing Through Music in Everyday Life*, 67.

53 Ibid., 22.

54 I referred to the placebo effect in Coyne, *Mood and Mobility: Navigating the Emotional Spaces of Digital Social Networks.*

55 Patrik N. Juslin and Daniel Västfjäll, 'Emotional Responses to Music: The Need to Consider Underlying Mechanisms', *Behavioural and Brain Sciences* 31 (2008): 559–621

56 Anderson, 'Transient Convergence and Relational Sensibility: Beyond the Modern Constitution of Nature'.

57 Vitruvius, *Vitruvius: The Ten Books on Architecture.*

58 There are many studies into walking as a cultural and political phenomenon. See Jean-François Augoyard, *Step by Step: Everyday Walks in a French Urban Housing Project* (Minneapolis: University of Minnesota Press, 2007); Francesco Careri, *Walkscapes: Walking as an Aesthetic Practice* Gustavo Gili, 2001). Colleague Katerina Talianni has examined 'sound walks', exploring the sonic aspects of outdoor walking. She references Ai Weiwei and Anish Kapoor's walk in London for refugees Martin Godwin, 'Ai Weiwei and Anish Kapoor Lead London Walk of Compassion for Refugees'. *Guardian*, 17 September (2015).

Available online: https://www.theguardian.com/uk-news/2015/sep/17/ai-weiwei-anish-kapoor-london-walk-refugees (accessed 3 June 2017); Bill Atkinson's 'Tour of all Tours' in which he takes people on group city tours exploring tours and tourism http://tourofalltours.blogspot.co.uk (accessed 3 June 2017). The group in Italy known as Stalker takes groups on 'conscious raising' treks to marginalized urban areas, such as Roma camps http://www.osservatorionomade.net (accessed 3 June 2017); Marina Abramović and her partner Ulay walked the length of the Great Wall of China from opposite ends, met in the middle and then officially broke off their romantic relationship 'Lovers Abramović & Ulay walk the Length of the Great Wall of China from Opposite Ends, Meet in the Middle and BreakUp', *Kickass Trips*, 14 January (2015). Available online: http://kickasstrips.com/2015/01/lovers-abramovic-ulay-walk-the-length-of-the-great-wall-of-china-from-opposite-ends-meet-in-the-middle-and-breakup/ (accessed 3 June 2017).

59 'Walking "Cuts Risk of Stroke in Men"', *Evening Standard*, 14 November (2013). Available online: http://www.standard.co.uk/news/health/walking-cuts-risk-of-stroke-in-men-8940563.html (accessed 3 June 2017).

60 Aristotle, *Metaphysics* (Adelaide, Australia: eBooks@Adelaide 2013), https://ebooks.adelaide.edu.au/a/aristotle/metaphysics/book5.html

61 Phenomenologist Erwin W. Straus (1891–1975) provides a succinct description of the mechanics of walking: 'Human gait, is in fact, a continuously arrested falling' (Erwin Walter Maximilian Straus, *Phenomenological Psychology: The Selected Papers of Erwin W. Straus* (London: Tavistock, 1966), 148).

62 On the four causes see Heidegger, *The Question Concerning Technology and Other Essays*, 6.

63 Augoyard, *Step by Step: Everyday Walks in a French Urban Housing Project*.

64 Ibid.

65 Menno Hulswit, 'Peirce on Causality and Causation'. *The Commens Encyclopedia: The Digital Encyclopedia of Peirce Studies. New Edition*, (2001), Pub. 120809-1715a. Available online: http://www.commens.org/encyclopedia/article/hulswit-menno-peirce-causality-and-causation (accessed 11 June 2017).

66 Peirce, 'The Neglected Argument for the Reality of God'.

67 Ibid., 436.

68 Ibid.

69 Kaplan, 'The Restorative Benefits of Nature: Toward an Integrative Framework'.

70 R. S. Ulrich, et al., 'Stress Recovery During Exposure to Natural and Urban Environments', ibid., 11 (1991): 201–30, 205.

71 Richard Louv, *Last Child in the Woods: Saving Our Children from Nature-Deficit Disorder* (London: Atlantic Books, 2005).

72 Ibid.

73 Ibid.

74 Houser, 'Introduction', xxxii.

75 Nancy M. Wells and Kristi S. Lekies, 'Nature and the Life Course: Pathways from Childhood Nature Experiences to Adult Environmentalism', *Children, Youth and Environments* 16, no. 1 (2006): 1–24.

76 Perez-Gomez, *Attunement: Architectural Meaning After the Crisis of Modern Science*, 2–3.

77 Hobbes, *Leviathan*.

78 Herbert Marcuse, *Eros and Civilisation: A Philosophical Inquiry into Freud* (London: Routledge and Kegan Paul, 1987), 166.

79 The CEO of Facebook announced a plan to engineer a disease-free future. See Mark Zuckerberg, 'Can we Cure all Diseases in Our Children's lifetime?'. *Facebook Notes*, 21 September (2016). Available online: https://www.facebook.com/notes/mark-zuckerberg/can-we-cure-all-diseases-in-our-childrens-lifetime/10154087783966634 (accessed 4 June 2017).

80 Isaiah 11:6.

81 I followed this line of argument in Coyne, *The Tuning of Place: Sociable Spaces and Pervasive Digital Media*.

82 Crawford Stanley Holling, 'Resiliency and Stability of Ecological Systems', *Annual Review of Ecology and Systematics* 4 (1973): 1–23, 15.

83 Ibid.

84 Friedrich Wilhelm Nietzsche, *The Nietzsche Reader* (London: Penguin, 1977), 281.

85 Anderson, 'Transient Convergence and Relational Sensibility: Beyond the Modern Constitution of Nature'.

86 DeNora, *Music Asylums: Wellbeing Through Music in Everyday Life*, 22.

Chapter 10

1 Christopher Street, *London's Ley Lines: Pathways of Enlightenment* (Earthstars Publishing, 2010), 3.

2 Ibid. A 'ley' is simply a pasture. For more about the land, see Chapter 6.

3 *OED*

4 Alfred Watkins, *Early British Trackways* (London: Simkin, Marshall, Hamilton, Kent and Co, 1922), 30–2.

5 Ibid.

6 Ibid.

7 Ibid.

8 Ibid.

9 Feng shui and geomancy are interesting in this respect. See Emile Durkheim and Marcel Maus, *Primitive Classification* (Chicago, IL: The University of Chicago Press, 1963).

10 See, for example, *The One Truth* website and item #12 at http://jandeane81.com/threads/8111-Rosslyn-Chapel (accessed 4 June 2017). The Centrality of Chartres seems to depend on which mapping projection is used.

11 Such ideas feature in speculative literature, fiction and of course the web: for example, Daniel Winter (2000), 'Ports of Gaol: "Eggs Files" for Dragons Feeling for Europe's Navigators – Spiritual Destiny of Port-U-Graal, Europe and the Great Dragon Line'. *ImplosionGroup*, August. Available online: http://www.bibliotecapleyades.net/sumer_anunnaki/reptiles/reptiles29.htm (accessed 11 June 2017).

12 On mood and place, see Coyne, *Mood and Mobility: Navigating the Emotional Spaces of Digital Social Networks*.

13 Peirce, 'Immortality in the Light of Synechism', 3.

14 Ibid.

15 Ibid.

16 Ibid.

17 Ibid.

18 Ibid., 1.

19 According to philosopher Bertrand Russell, monism is the doctrine 'that the world as a whole is a single substance, none of whose parts are logically capable of existing alone' (Bertrand Russell, *A History of Western Philosophy* (New York, NY: Simon and Schuster, 1945), 577). Monism challenges the mind–body dualism of Rene Descartes's philosophy, and other oppositions.

20 John D. Caputo, *On Religion* (London: Routledge, 2001), 87.

21 George Lucas, 'Yoda Quotes'. *Genius Album Guides* (2017). Available online: https://genius.com/George-lucas-yoda-quotes-annotated (accessed 4 June 2017).

22 ricca_riot. (2015), 'Interstellar Transmissions'. *Archive of Our Own*, Chapter 2. Available online: http://archiveofourown.org/works/5496170/chapters/12742442?view_adult=true (accessed 4 June 2017).

23 Rupert Sheldrake, *Morphic Resonance: The Nature of Formative Causation* (Rochester, Vermont: Park Street Press, 2009), xxv.

24 *OED*

25 Gaétan Chevalier, et al., 'Earthing: Health Implications of Reconnecting the Human Body to the Earth's Surface Electrons', *Journal of Environmental and Public Health*, no. doi: 10.1155/2012/291541 (2012): 1–23.

26 Ibid., 2.

27 Ibid.

28 The 1968 film production of *A Midsummer Night's Dream* by Peter Hall is available on Youtube at https://www.youtube.com/watch?v=4RD-7aRcxmA. This complete version presents as a hippy woodland semi-nude romp.

29 Lou Agnes Reynolds and Paul Sawyer, 'Folk Medicine and the Four Fairies of *A Midsummer Night's Dream*', *Shakespeare Quarterly* 10, no. 4 (1959): 513–21, 517.

30 Coyne, *Technoromanticism: Digital Narrative, Holism, and the Romance of the Real.*

31 Kevin Robins, 'Cyberspace and the World We Live In', in *Cyberspace, Cyberbodies, Cyberpunk: Cultures of Technological Embodiment*, ed. M. Featherstone and R. Burrows, (London: Sage), 135–56, 139.

32 Marcos Novak, 'Liquid Architectures in Cyberspace', in *Cyberspace: First Steps*, ed. M. Benedikt (Cambridge, MA: MIT Press), 225–54, 226. Also see Stephen O'Leary, 'Cyberspace as Sacred Space: Communicating Religion on Computer Networks', in *Religion Online: Finding Faith on the Internet*, ed. L. L. Dawson and D. E. Cowan,37–58.

33 Frank L. Baum, *The Wonderful Wizard of Oz* (Oxford: Oxford University Press, 1997).

34 Susan Wolstenholme, 'Introduction', in *The Wonderful Wizard of Oz by Frank L. Baum*, ed. S. Wolstenholme (Oxford: Oxford University Press), ix–xliii, xxxvi.

35 T. L. S. Sprigge, 'Pantheism', *The Monist* 80, no. 2 (1997): 191–217, 193.

36 Ibid.

37 William Indick, *The Digital God: How Technology Will Reshape Spirituality* (Jefferson, NC: McFarland, 2015).

38 Ibid., 35.

39 Ibid.

40 Ibid.

41 Ibid.

42 Ibid., 35–6.

43 Ibid.

44 Erik Davis, 'Technopagans: May the Astral Plane be Born in Cyberspace'. *Wired Magazine*, 1 July (1995). Available online: http://www.wired.com/1995/07/technopagans/ (accessed 11 June 2017).

45 Huizinga, *Homo Ludens: A Study of the Play Element in Culture*, 10.

46 Jesper Juul, 'The Magic Circle and the Puzzle Piece', in *Conference Proceedings of the Philosophy of Computer Games 2008*, ed. S. Günzel, M. Liebe and D. Mersch (Potsdam: Potsdam University Press), 56–67.

47 Joshua Fairfield, 'The Magic Circle', *Vanderbilt Journal of Entertainment and Technology Law*, Washington & Lee Legal Studies Paper No. 2008-45, no. (2009):

48 MSc by research student Yao Zhang alerted me to the theme of 'reset' in film and video games. In an essay, she also references the film *Run Lola Run* and the *Scott Pilgrim Versus the World* graphic novels and film. The reset idea has crossed over from video games to films and political commentary: Fiona Parker, 'Brexit Terms Should be "Reset," Labour's Shadow Brexit Secretary Says'. *Metro*, 9 June (2017). Available online: http://metro.co.uk/2017/06/09/brexit-terms-should-be-reset-labours-shadow-brexit-secretary-says-6696996/ (accessed 10 June 2017).

49 https://www.youtube.com/watch?v=0v2xnl6LwJE.

50 http://www.monumentvalleygame.com.

51 According to philosopher David Ray Griffin, the mechanistic philosophy of Rene Descartes helped counteract the belief that minds can control objects over distance, eventually reducing the persecution of people branded as witches. See David Ray Griffin, 'Introduction: The Reenchantment of Science', in *The Reenchantment of Science: Postmodern Proposals*, ed. D. R. Griffin (Albany, NY: SUNY Press), 1–46, 2.

52 Virginia Hefferman, *Magic and Loss: The Internet as Art* (London: Simon and Schuster, 2016), 17. In Russell T Davies' BBC adaptation of *A Midsummer Night's Dream* in 2016 the attendants in the court of the Duke of Athens consulted tablet computers, and Lysander and Hermia used a wall-mounted digital display screen to plan their escape to the forest.

53 Ibid.

54 Peirce, 'What is a Sign?', 9.

55 Mercea Eliade, *Images and Symbols: Studies in Religious Symbolism* (London: Harvill Press, 1961), 44.

56 Ibid., 83.

57 Max Weber, 'Science as a Vocation', in *Max Weber: Essays in Sociology*, ed. H. H. Gerth and C. Wright Mills (New York: Oxford University Press), 129–56, 155.

58 Griffin, 'Introduction: The Reenchantment of Science', 2.

59 Peter Pels, 'Introduction', in *Magic and Modernity: Interfaces of Revelation and Concealment*, ed. B. Meyer and P. Pels (Stanford, CA: Stanford University Press), 1–38, 4.

60 Martin Heidegger, 'Building, Dwelling, Thinking', in *Poetry, Language, Thought* (New York: Harper and Rowe), 143–61, 171.

61 Ibid., 143.

62 Jacques Derrida, 'Plato's Pharmacy', in *Dissemination* (London: Athlone), 61–171.

63 Oliver Burkeman, 'Five Reasons why we Should all Learn how to do Nothing'. *Guardian*, 4 June (2015). Available online: https://www.theguardian.com/lifeandstyle/2015/jan/09/five-reasons-we-should-all-learn-to-do-nothing (accessed 9 January).

64 Espen Hammer, 'Being Bored: Heidegger on Patience and Melancholy', *British Journal for the History of Philosophy* 12, no. 2 (2004): 277–95, 283.

65 Ibid., 285.

66 Martin Heidegger, *The Fundamental Concepts of Metaphysics: World, Finitude, Solitude* (Bloomington, IN: Indiana University Press, 1995), 153.

67 Elizabeth S. Goodstein, *Experience Without Qualities: Boredom and Modernity* (Stanford, CA: Stanford University Press, 2005), 325.

Chapter 11

1 Consider for example self-management and autarky via community gardening as tools for 'citizen participation and empowerment' (Kaminer, *The Efficacy of Architecture: Political Contestation and Agency,* 108).

2 Emerson, *Essays by Ralph Waldo Emerson.*

3 Martin Heidegger, 'The Principle of Ground', *Man and World* 7, no. (1974): 207–22.

4 Emerson, *Essays by Ralph Waldo Emerson,* 39.

5 The Outward Bound website lists training in self-reliance among the skill set it imparts to leaders in training: https://www.outwardbound.org.uk (accessed 29 October 2016).

6 Andrew Ballantyne, 'The Unit of Survival', *Arq (Architectural Research Quarterley)* 20, no. 1 (2016): 39–44, 43.

7 Bryan K. Saville, et al., 'Internet Addiction and Delay Discounting in College Students', *The Psychological Record* 60, no. (2010): 273–86; Ofcom, *The Communications Market Report.*

8 H. Bouma, et al., 'Gerontechnology in Perspective', *Gerontechnology* 6, no. 4 (2007): 190–216.

9 Nicholas Negroponte, *Being Digital* (London: Hodder and Stoughton, 1995), 6. Also see a critique by Philip Leonard, *Literature After Globalization: Textuality, Technology and the Nation-State* (London: Bloomsbury, 2013).

10 Nicholas Negroponte, 'Beyond Digital', *Wired Magazine* 6, no. 12 (1998):

11 Though plastic is undergoing a cultural revival considering its use a medium for 3D printing.

12 Kim Cascone, 'The Aesthetics of Failure: "Post-digital" Tendencies in Contemporary Computer Music', *Computer Music Journal* 24, no. 4 (2000): 12–18.

13 Ibid., 16.

14 I explore interference and the glitch in Coyne, *The Tuning of Place: Sociable Spaces and Pervasive Digital Media.*

15 David M. Berry and Michael Dieter, 'Thinking Postdigital Aesthetics: Art, Computation and Design', in *Postdigital Aesthetics: Art, Computation and Design*, ed. D. M. Berry and M. Dieter (Basingstoke, England: Palgrave Macmillan), 1–11; David M. Berry and Michael Dieter (eds), *Postdigital Aesthetics: Art, Computation and Design* (Basingstoke, England: Palgrave Macmillan, 2015); Florian Cramer, 'What is "Post-Digital"?', in *Postdigital Aesthetics: Art, Computation and Design*, ed. D. M. Berry and M. Dieter (Basingstoke, England: Palgrave Macmillan), 12–26.

16 Florian Cramer, 'What is "Post-digital"?', *A Peer-Reviewed Journal About Post-Digital Research* (2015): ; Cramer, 'What is "Post-digital"?'

17 Cramer, 'What is "Post-digital"?', 13.

18 https://www.youtube.com/watch?v=8Ck2q3YgRlY (accessed 14 March 2017).

19 Cramer, 'What is "Post-digital"?', 7.

20 Ibid., 20

21 Ibid., 25.

22 Ibid., 24. For photographs of prominent writers and their typewriters, see Marta Bausells 'Typewriters and Their Owners: Famous Authors at Work – in Pictures'. *Guardian*, 5 November (2014). Available online: https://www.theguardian.com/books/booksblog/gallery/2014/nov/05/typewriters-and-their-owners-famous-authors-at-work-in-pictures?CMP=share_btn_link (accessed 4 June 2017).

23 Cramer, 'What is "Post-digital"?', 22.

24 David M. Berry and Michael Dieter, 'Thinking Postdigital Aesthetics: Art, Computation and Design', ibid., 1–11.

25 Florian Cramer, 'What is "Post-digital"?', ibid., 12–26, 23.

26 Ibid.

27 Ibid., 15.

28 Ibid., 13.

29 For a seductive catalogue of outdoor accessories and nature settings, see Robert Klanten, et al., *The Outsiders: New Outdoor Creativity* (Berlin: Gestalten, 2014).

30 An instructional video from the 1950s makes the point. See https://www.youtube.com/watch?v=rQ6EwJFLJ34 (accessed 10 June 2017).

31 Clay-Atkinson 'A Beginner's Guide to Preparedness'. *Self-Reliance*, 2 October (2014). Available online: http://www.self-reliance.com/2014/10/a-beginners-guide-to-preparedness/ (accessed 11 June 2017).

32 Ibid.

33 Ibid.

34 Allen Mendenhall, 'The Classical Liberalism of Ralph Waldo Emerson'. *The Literary Lawyer: A Forum for the Legal and Literary Communities*, 7 January (2015). Available online: https://allenmendenhallblog.com/2015/01/07/the-classical-liberalism-of-ralph-waldo-emerson/ (accessed 11 June 2017).

35 Ronald Tiersky, 'Ralph Waldo Emerson and Donald Trump: Will Power Educate the Potentate?'. *Huffington Post*, 17 May (2016). Available online: http://www.huffingtonpost.com/ronald-tiersky/ralph-waldo-emerson-and-d_b_9998844.html (accessed 11 June 2017).

36 Emerson, *Essays by Ralph Waldo Emerson*, 35.

37 Thoreau, *Walden: Or, Life in the Woods.*

38 Gary North, 'Thoreau's Walden: Phony Testament of the Greens'. *Gary North's Specific Answers*, 18 April (2014). Available online: https://www.garynorth.com/public/12347.cfm (accessed 4 June 2017).

39 Emerson, *Essays by Ralph Waldo Emerson*, 34.

40 David Golumbia, 'Bitcoin as Politics: Distributed Right-wing Extremism', in *Moneylab Reader: An Intervention in Digital Economy*, ed. G. Lovink, N. Tkacz and P. de Vries (Amsterdam: Institute of Network Cultures), 117–31, 119.

41 I am Legend (2007).

42 John Wyndham, *The Day of the Triffids* (London: Penguin, 1971).

43 Planet of the Apes (1968).

44 Karl Marx, 'Capital', in *Karl Marx: Selected Writings*, ed. D. McLellan (Oxford: Oxford University Press), 415–507.

45 Daniel C. Dennett, 'Where am I?', in *Brainstorms: Philosophical Essays on Mind and Psychology*, ed. D. C. Dennett (Cambridge, MA: MIT Press), 310–23; Terrel Miedaner, 'The Soul of the Mark III Beast (and Reflection)', in *The Mind's Eye: Fantasies and Reflections on Self and Soul*, ed. D. R. Hofstadter and D. C. Dennett (New York: Basic Books), 109–15.

46 As in the film, *The Man with Two Brains* (1983). Also see Charlie Gere, 'Brains-in-vats, Giant Brains and World Brains: The Brain as Metaphor in Digital Culture', *Studies in History and Philosophy of Biological and Biomedical Sciences* 35, no. (2004): 351–66.

47 Moravec, *Mind Children: The Future of Robot and Human Intelligence*.

48 Amanda Gefter, 'The Evolutionay Argument Against Reality'. *Quanta Magazine*, 21 April (2016). Available online: https://www.quantamagazine.org/20160421-the-evolutionary-argument-against-reality/ (accessed 11 June 2017).

49 I'm drawn to the observation by Humberto Maturana and Francisco Varela that a frog sees mainly a world of flies, and is attuned to their detection and capture. See Humberto Maturana, 'Reality: The Search for Objectivity or the Quest for a Compelling Argument', *The Irish Journal of Psychology* 9, no. 1 (1988): 25–82; Humberto Maturana and Francisco G. Varela, *Autopoiesis and Cognition* (Dordrecht: Reidel, 1980), and an account of their theories of autopoesis by Terry Winograd and Fernando Flores, *Understanding Computers and Cognition: A New Foundation for Design* (Reading, MA: Addison-Wesley, 1986).

50 Nick Bostrom, *Superintelligence: Paths, Dangers, Strategies* (Oxford, UK: Oxford University Press, 2014); Nick Bostrom, 'Are You Living in a Simulation?', *Philosophical Quarterly* 53, no. 211 (2003): 243–55.

51 Bostrom, 'Are You Living in a Simulation?', 5.

52 Based on the book by Philip K. Dick, *We Can Remember It for You Wholesale* (London: Millennium, 2000).

53 Plotinus, *The Essence of Plotinus: Extracts from the Six Enneads and Porphyry's Life of Plotinus* (New York: Oxford University Press, 1948).

54 George Berkeley, *A Treatise Concerning the Principles of Human Knowledge* (Philadelphia, PA: J.B. Lippincott, 1874).

55 Edward Hallett Carr, *What is History?* (Harmondsworth: Penguin, 1964).

56 For example, see Sabine Hessenfelder, 'The Simulation Hypothesis and other Things I don't Believe'. *BackReaction Philosophy Blog*, 28 February (2013). Available online: http://backreaction.blogspot.co.uk/2013/02/the-simulation-hypothesis-and-other.html (accessed 11 June 2017).

57 Bostrom, 'Are You Living in a Simulation?', 2.

58 Maturana and Varela, *Autopoiesis and Cognition*; Hutchins, *Cognition in the Wild*; Andy Clark and David Chalmers, 'The Extended Mind', *Analysis* 58, no. 1 (1998): 7–19.

59 E. M. Forster, *The Machine Stops* (London: Penguin, 2011).

60 Mark Johnson, *The Body in the Mind: The Bodily Basis of Meaning, Imagination, and Reason* (Chicago, IL: University of Chicago Press, 1987); George Lakoff, *Women, Fire, and Dangerous Things: What Categories Reveal about the Mind* (Chicago: University of Chicago Press, 2003); George Lakoff and Mark Johnson, *Metaphors We Live By* (Chicago, IL: University of Chicago Press, 1980).

61 The 'equipmental whole' is a term developed by Martin Heidegger in Heidegger, *Being and Time*. Also see Dreyfus, *Being-in-the-world: A Commentary on Heidegger's Being and Time, Division I.*

62 Heidegger, *Being and Time.*

63 Elizabeth Halsted, 'A Shoe is Rarely Just a Shoe: Women's Accessories and their Psyches', in *Longing: Psychoanalytic Musings on Desire*, ed. J. Petrucelli (London: Karnac), 101–11, 110.

64 Tetsuro Watsuji, *Watsuji Tetsuro's Rinrigaku: Ethics in Japan* (Albany: State University of New York Press, 1996), 22. He champions the idea of negation. Of nature, he says: 'At issue here is the natural world, which arises in an intersubjective way; and hence, it arises within consciousness in general' (p. 179).

65 Ibid.

66 Easley and Kleinberg, *Networks, Crowds and Markets: Reasoning About a Highly Connected World,* 1. For a compelling account of the potential causes and consequences of power blackouts, see Hugh Byrd and Steve Matthewman, 'Exergy and the City: The Technology and Sociology of Power (failure)', *Journal of Urban Technology* 21, no. 3 (2014): 85–102.

67 The Twitter streams that emerged during the 2016 US presidential campaign and its aftermath provide a case in point.

Coda

1 Lakoff and Johnson, *Metaphors We Live By.*

2 Consider the mood of melancholy, that entertains the possibility of feeling sad while also feeling happy, and other complexities within emotion and affect. See Coyne, *Mood and Mobility: Navigating the Emotional Spaces of Digital Social Networks.*

3 Churchman, 'Wicked Problems'; Rittel and Webber, 'Dilemmas in a General Theory of Planning'; Buchanan, 'Wicked Problems in Design Thinking'; Coyne, 'Wicked Problems Revisited'

4 (The Elder) Pliny, *Natural History: A Selection* (London: Penguin, 1991), 326.

5 Ibid., 330.

6 Mitchell, *Picture Theory: Essays on Verbal and Visual Representation.*

REFERENCES

Abel, C. (2015), *The Extended Self: Architecture, Memes and Minds*, Manchester, England: Manchester University Press.

Abel, C. (2017), *Architecture and Identity: Responses to Cultural and Technological Change*, Abingdon, Oxon: Routledge.

Abram, D. (2011), *Becoming Animal: An Earthly Cosmology*, New York: Vintage.

Agamben, G. (2004), *The Open: Man and Animal*, Stanford, CA: Stanford University Press.

Alberti, L. B. (1996), *On the Art of Building in Ten Books*, trans. J. Rykwert, N. Leach and R. Tavernor, Cambridge, MA: MIT Press.

Albrecht, G. (2005), 'Solastalgia: A New Concept in Health and Identity', *PAN: Philosophy, Activism, Nature*, 3: 44–59.

Alexander, C. (1988), 'A City Is Not a Tree', in J. Thackara (ed.), *Design after Modernism*, 67–84, London: Thames and Hudson.

Andersen, P. B. (1992), 'Computer Semiotics', *Scandinavian Journal of Information Systems*, 4 (1): 3–30.

Anderson, C. (2008), 'The End of Theory: The Data Deluge Makes the Scientific Method Obsolete'. *Wired Magazine*, 23 June. Available online: http://archive.wired.com/science/discoveries/magazine/16-07/pb_theory (accessed 11 June 2017).

Anderson, D. R. (1984), *Creativity and the Philosophy of C.S. Peirce (Abduction, Evolution, God)*, Ann Arbor, MI: PhD Thesis, Pennsylvania State University.

Anderson, D. R. (1987), *Creativity and the Philosophy of C.S. Peirce*, Dordrecht, The Netherlands: Springer.

Anderson, J. (2009), 'Transient Convergence and Relational Sensibility: Beyond the Modern Constitution of Nature', *Emotion, Space and Society*, 2: 120–7.

Anderson, N. (2011), 'Lulzsec Manifesto: "We Screw Each Other over for a Jolt of Satisfaction"'. *Ars Technica*, 17 June. Available online: https://arstechnica.com/tech-policy/2011/06/lulzsec-heres-why-we-hack-you-bitches/ (accessed 3 June 2017).

Ansted, D. T. (1863), *The Great Stone Book of Nature*, London: Macmillan & Co.

Appleton, J. (1975), *The Experience of Landscape*, London: John Wiley.

Appleton, J. (1984), 'Prospects and Refuges Re-Visited', *Landscape Journal*, 3 (2): 91–103.

Appleyard, D., K. Lynch and J. R. Myer. (1966), *The View from the Road*, Cambridge, MA: Joint Center for Urban Studies of the Massachusetts Institute of Technology and Harvard University.

Aristotle. (2013), *Metaphysics*, trans. W. D. Ross, Adelaide, Australia: eBooks@Adelaide

Armstrong, R. (2015), *Vibrant Architecture: Matter as a Codesigner of Living Structures*, Warsaw: Gruyter Open.

Aspinall, P., P. Mavros, R. Coyne and J. Roe. (2013), 'The Urban Brain: Analysing Outdoor Physical Activity with Mobile Eeg', *British Journal of Sports Medicine*, 49 (4): http://dx.doi.org/10.1136/bjsports-2012-091877.

Atkin, A. (2005), 'Peirce on the Index and Indexical Reference', *Transaction of the Charles S. Peirce Society*, 41 (1): 161–88.

Atkinson, M. (2009), 'Parkour, Anarcho-Environmentalism, and Poiesis', *Journal of Sport and Social Issues*, 33 (2): 169–94.

Augé, M. (1995), *Non-Places: Introduction to an Anthropology of Supermodernity*, trans. J. Howe, London: Verso.

Augoyard, J.-F. (2007), *Step by Step: Everyday Walks in a French Urban Housing Project*, trans. D. A. Curtis, Minneapolis: University of Minnesota Press.

Aurelius, M. (1964), *Meditations*, trans. M. Staniforth, London: Penguin.

Aurigi, A. (2016), 'No Need to Fix: Strategic Inclusivity in Developing and Managing the Smart City', in G. A. Caldwell, S. C. H. and E. M. Clift (eds), *Digital Futures and the City of Today: New Technologies and Physical Spaces*: 9–27, Bristol: Intellect.

Baker, V. R. (1999), 'Geosemiosis', *Bulletin of the Geological Society of America*, 111 (5): 633–45.

Balding, R. (2012), 'Turn Off Your Smartphone to Beat Stress'. *British Psychological Society*, 13 June. Available online: http://wp.me/p2tvLx-cz (accessed 11 June 2017).

Ballantyne, A. (2016), 'The Unit of Survival', *Arq (Architectural Research Quarterley)*, 20 (1): 39–44.

Banos, O., et al. (2014), 'mHealthdroid: A Novel Framework for Agile Dev. Of Mobile Health Apps', in L. Pecchia, L. L. Chen, C. Nugent and J. Bravo (eds), *Lecture Notes in Computer Science, Vol 8868*: 91–8, Cham: Springer.

Barbieri, M. e. (2008), *Introduction to Biosemiotics: The New Biological Synthesis*, Dordrecht, the Netherlands: Springer.

Barthes, R. (1973), *Mythologies*, trans. A. Lavers, London: Paladin.

Barthes, R. (1977), 'The Death of the Author', in S. Heath (ed), *Image, Music, Text*: 142–9, London: Fontana.

Bastian, M. (2012), 'Fatally Confused: Telling the Time in the Midst of Ecological Crises', *Environmental Philosophy*, 9 (1): 23–48.

Bastian, M. and T. van Dooren. (2017), 'The New Immortals: Immortality and Infinitude in the Anthropocene', *Journal of Environmental Philosophy*, 14 (1): 1–9.

Battail, G. (2009), 'Applying Semiotics and Information Theory to Biology: A Critical Comparison', *Biosemiotics*, 2: 303–20.

Baudrillard, J. (1994), *Simulacra and Simulation*, trans. S. F. Glaser, Ann Arbor: University of Michigan press.

Baum, F. L. (1997), *The Wonderful Wizard of Oz*, Oxford: Oxford University Press.

Bausells, M. (2014), 'Typewriters and Their Owners: Famous Authors at Work – in Pictures'. *Guardian*, 5 November. Available online: https://www.theguardian.com/books/booksblog/gallery/2014/nov/05/typewriters-and-their-owners-famous-authors-at-work-in-pictures?CMP=share_btn_link (accessed 4 June 2017).

Beatley, T. and P. Newman. (2013), 'Biophilic Cities Are Sustainable, Resilient Cities', *Sustainability*, 5: 3328–45.

Begby, E. (2014), 'Hermeneutics and Pragmatism', in J. Malpas and H.-H. Gander (eds), *The Routledge Companion to Hermeneutics*: 612–22, Abingdon, Oxon: Routledge.

Benjamin, W. (2003), *The Origin of German Tragic Drama*, trans. G. Steiner and J. Osborne, London: Verso.

Berger, J. (1972), *Ways of Seeing*, London: Penguin.

Berger, J. (1977), 'Why Zoos Disappoint', *New Society*, 40: 122–3.

Berkeley, G. (1874), *A Treatise Concerning the Principles of Human Knowledge*, Philadelphia, PA: J.B. Lippincott.

Berry, D. M. and M. Dieter (eds). (2015), *Postdigital Aesthetics: Art, Computation and Design*, Basingstoke, England: Palgrave Macmillan.

Berry, D. M. and M. Dieter. (2015), 'Thinking Postdigital Aesthetics: Art, Computation and Design', in D. M. Berry and M. Dieter (eds), *Postdigital Aesthetics: Art, Computation and Design*: 1–11, Basingstoke, England: Palgrave Macmillan.

Bertalanffy, L. (1969), *General System Theory: Foundations, Development, Applications*, New York: G. Braziller.

Black, M. (1962), *Models and Metaphors: Studies in Language and Philosophy*, Ithaca: Cornell University Press.

Blackmore, S. (1999), *The Meme Machine*, Oxford, England: Oxford University Press.

Blackwell, A. (2013), *Visit Sunny Chernobyl: Adventures in the World's Most Polluted Places*, London: Random.

'Blockchains: The Great Chain of Being Sure About Things'. (2015), *The Economist*, 31 October. Available online: http://www.economist.com/news/briefing/21677228-technology-behind-bitcoin-lets-people-who-do-not-know-or-trust-each-other-build-dependable (accessed 15 June 2017).

Bloomer, K. (2008), 'The Picture Window: The Problem of Viewing Nature through Glass', in S. R. Kellert, J. Heerwagen and M. Mador (eds), *Biophilic Design: The Theory, Science and Practice of Bringing Buildings to Life*: 253–62, Hoboken, NJ: Wiley.

Bostrom, N. (2003), 'Are You Living in a Simulation?', *Philosophical Quarterly*, 53 (211): 243–55.

Bostrom, N. (2014), *Superintelligence: Paths, Dangers, Strategies*, Oxford, UK: Oxford University Press.

Botar, O. (2016), 'The Biocentric Bauhaus', in C. N. Terranova and M. Tromble (eds), *The Routledge Companion to Biology in Art and Architecture*: 17–51, London: Routledge.

Bouma, H., J. L. Fozard, D. G. Bouwhuis and V. T. Taipale. (2007), 'Gerontechnology in Perspective', *Gerontechnology*, 6 (4): 190–216.

Bratton, B. H. (2015), *The Stack: On Software and Sovereignty*, Cambridge, MA: MIT Press.

Brenton, H., M. Gillies, D. Ballin and D. Chatting. (2005), 'The Uncanny Valley: Does It Exist?' *Proceedings of Conference of Human Computer Interaction, Workshop on Human Animated Character Interaction*.

Broadbent, G. (1973), *Design in Architecture: Architecture and the Human Sciences*, New York, NY: John Wiley and Son.

Broadbent, G. (1980), 'Building Design as an Iconic Sign System', in G. Broadbent, R. Bunt and C. Jencks (eds), *Signs, Symbols, and Architecture*: 311–31, Chichester, England: John Wiley and Sons.

Broadbent, G. and J. Glusberg (ed.). (1991), *Deconstruction: A Student Guide*, London:

Brown, M., et al. (2016), 'Gamification and Adherence to Web-Based Mental Health Interventions: A Systematic Review', *JMIR Mental Health*, 3 (3).

Buchanan, R. (1995), 'Wicked Problems in Design Thinking', in V. Margolin and R. Buchanan (eds), *The Idea of Design*: 3–20, Cambridge, MA: MIT Press.

Buchanan, R. (2001), 'Children of the Moving Present: The Ecology of Culture and the Search for Causes in Design', *Design Issues*, 17 (1): 67–84.

Budashevskaya, O. and B. Brownell. (2015), *Hypernatural: Architecture's New Relationship with Nature*, Princeton, NJ: Princeton University Press.

Bull, G. (2013), *Public Health and Landscape: Creating Healthy Places*, London: Landscape Institute.

Bullivant, L. e. (2017), *4d Hyper-Local: A Cultural Tool Kit for the Open Source City*, Chichester, UK: Wiley.

Burke, E. and J. Boulton (ed.). (1958), *A Philosophical Enquiry into the Origin of Our Ideas of the Sublime and Beautiful*, Notre Dame: University of Notre Dame Press.

Burkeman, O. (2015), 'Five Reasons Why We Should All Learn How to Do Nothing'. *Guardian*, 4 June. Available online: https://www.theguardian.com/lifeandstyle/2015/jan/09/five-reasons-we-should-all-learn-to-do-nothing (accessed 9 January).

Burnett-Zeigler, I. E. (2016), 'How Donald Trump Affects Therapy Patients'. *Time*, 28 November. Available online: http://time.com/4583628/donald-trump-therapy-patients/ (accessed 3 June 2017).

Butler, J. (1990), *Gender Trouble: Feminism and the Subversion of Identity*, New York: Routledge.

Byrd, H. and S. Matthewman. (2014), 'Exergy and the City: The Technology and Sociology of Power (Failure)', *Journal of Urban Technology*, 21 (3): 85–102.

Byrne, J. and N. Osborne. (2016), 'Urban Hacktivism: Getting Creative About Involving Citizens in City Planning'. *The Conversation (blog site)*, 5 July. Available online: https://theconversation.com/urban-hacktivism-getting-creative-about-involving-citizens-in-city-planning-62277 (accessed 3 June 2017).

Cadwalladr, C. (2017), 'Robert Mercer: The Big Data Billionaire Waging War on Mainstream Media'. *Guardian*, 26 February. Available online: https://www.theguardian.com/politics/2017/feb/26/robert-mercer-breitbart-war-on-media-steve-bannon-donald-trump-nigel-farage (accessed 11 June 2017).

Caillois, R. (1961), *Man, Play, and Games*, New York, NY: The Free Press of Glencoe.

Caillois, R. (1984), 'Mimicry and Legendary Psychasthenia', *October*, 31 (Winter): 17–32 (First published in *Minotaure* in 1935).

Caldecott, R. (2006), 'The Babes in the Wood'. *Project Gutenberg EBook*. Available online: http://www.gutenberg.org/files/19361/19361-h/19361-h.htm (accessed 11 June 2017).

Caputo, J. D. (2001), *On Religion*, London: Routledge.

Careri, F. (2001), *Walkscapes: Walking as an Aesthetic Practice*, Gustavo Gili.

Carr, E. H. (1964), *What Is History?*, Harmondsworth: Penguin.

Cascone, K. (2000), 'The Aesthetics of Failure: "Post-Digital" Tendencies in Contemporary Computer Music', *Computer Music Journal*, 24 (4): 12–8.

Castells, M. (1996), *The Rise of the Network Society*, Malden, MA: Blackwell Publishers.

Castells, M., M. Fernandez-Ardevol, J. L. Qiu and A. Sey. (2007), *Mobile Communication and Society: A Global Perspective*, Cambridge, MA: MIT Press.

Cernea, D., A. Kerren and A. Ebert. (2011), 'Detecting Insight and Emotion in Visualization Applications with a Commercial EEG Headset', *Proc. Sigrad* 2011: 53–60.

Chalmers, A. F. (1999), *What Is This Thing Called Science?*, Indianapolis, IN: Hackett Pub Co.

Chevalier, G., et al. (2012), 'Earthing: Health Implications of Reconnecting the Human Body to the Earth's Surface Electrons', *Journal of Environmental and Public Health* (doi: 10.1155/2012/291541): 1–23.

Chu-Carroll, M. (2012), 'Turing Machines: What They Are, What They Aren't'. *Good Math, Bad Math (A Scientopia blog)*, 24 June. Available online: http://goodmath.scientopia.org/2012/06/24/turing-machines-what-they-are-what-they-arent/ (accessed 11 June 2017).

Churchman, C. W. (1967), 'Wicked Problems', *Management Science*, 14 (4): B141–B2.

Clark, A. (1997), *Being There: Putting Brain, Body and World Together Again*, Cambridge, MA: MIT Press.

Clark, A. and D. Chalmers. (1998), 'The Extended Mind', *Analysis*, 58 (1): 7–19.

Clark, N. (2017), 'Pyropolitics for a Planet of Fire', in K. Peters, P. Steinberg and E. Stratford (eds), *Territory Beyond Terra*: to appear, London: Rowman and Littlefield.

Clark, T. (2015), *Ecocriticism on the Edge: The Anthropocene as a Threshold Concept*, London: Bloomsbury.

Clarke, P. (2000), 'Adam Smith, Stoicism and Religion in the 18th Century', *History of the Human Sciences*, 13 (4): 49–72.

Clay-Atkinson. (2014), 'A Beginner's Guide to Preparedness'. *Self-Reliance*, 2 October. Available online: http://www.self-reliance.com/2014/10/a-beginners-guide-to-preparedness/ (accessed 11 June 2017).

Commision, F. (2009), *Woods for Health Strategy*, Edinburgh: Forestry Commission Scotland.

Conlin, J. (2013), *The Pleasure Garden, from Vauxhall to Coney Island*, Philadelphia, PA: University of Pennsylvania Press.

Cowley, S. J. (2014), 'Bio-Ecology and Language: A Necessary Unity', *Language Sciences*, 41: 60–70.
Coyne, R. (1995), *Designing Information Technology in the Postmodern Age: From Method to Metaphor*, Cambridge, MA: MIT Press.
Coyne, R. (1999), *Technoromanticism: Digital Narrative, Holism, and the Romance of the Real*, Cambridge, MA: MIT Press.
Coyne, R. (2005), *Cornucopia Limited: Design and Dissent on the Internet*, Cambridge, MA: MIT Press.
Coyne, R. (2005), 'Wicked Problems Revisited', *Design Studies*, 26 (1): 5–17.
Coyne, R. (2008), 'The Net Effect: Design, the Rhizome, and Complex Philosophy', *Futures*, 40: 552–61.
Coyne, R. (2010), *The Tuning of Place: Sociable Spaces and Pervasive Digital Media*, Cambridge, MA: MIT Press.
Coyne, R. (2011), *Derrida for Architects*, Abingdon: Routledge.
Coyne, R. (2016), 'Brainwalks'. *Reflections on Technology, Media and Culture*, 18 March. Available online: https://richardcoyne.com/2017/03/18/brainwalks/ (accessed 14 June 2017).
Coyne, R. (2016), *Mood and Mobility: Navigating the Emotional Spaces of Digital Social Networks*, Cambridge, MA: MIT Press.
Coyne, R. (2017), 'Melancholy Urbanism: Distant Horizons and the Presentation of Place', in F. Penz and R. Koeck (eds), *Cinematic Urban Geographies*, London: Palgrave Macmillan.
Coyne, R., et al. (1990), *Knowledge-Based Design Systems*, Reading, MA: Addison-Wesley.
Coyne, R., A. Snodgrass and D. Martin. (1994), 'Metaphors in the Design Studio', *JAE (Journal of Architectural Education)*, 48 (2): 113–25.
Cramer, F. (2015), 'What Is "Post-Digital"?', *A Peer-Reviewed Journal About Post-Digital Research*.
Cramer, F. (2015), 'What Is "Post-Digital"?', in D. M. Berry and M. Dieter (eds), *Postdigital Aesthetics: Art, Computation and Design*: 12–26, Basingstoke, England: Palgrave Macmillan.
Crowe, S. (1958), *The Landscape of Power*, London: Architectural Press.
Cruz, M. (2013), *The Inhabitable Flesh of Architecture*, Farnham, Surrey, England: Ashgate.
Cruz, M. and R. Beckett. (2016), 'Bioreceptive Design: A Novel Approach to Bio-Digital Materiality', *Arq (Architectural Research Quarterley)*, 20 (1): 51–64.
Cullen, G. (1961), *Concise Townscape*, Abingdon, England: Architectural Press.
Darwin, C. (1996), *The Origin of Species*, Oxford: Oxford University Press.
Davidson, W. L. (1979), *The Stoic Creed*, New York: Arno Press.
Davies, J. A. (2016), 'Machines for Living In: Connections and Contrasts between Designed Architecture and the Development of Living Forms', *Arq (Architectural Research Quarterley)*, 20 (1): 45–50.
Davis, E. (1995), 'Technopagans: May the Astral Plane Be Born in Cyberspace'. *Wired Magazine*, 1 July. Available online: http://www.wired.com/1995/07/technopagans/ (accessed 11 June 2017).
Dawkins, R. (1976), *The Selfish Gene*, New York: Oxford University Press.
Dawkins, R. (1996), *The Blind Watchmaker: Why the Evidence of Evolution Reveals a Universe without Design*, New York: Norton.
Deleuze, G. and F. Guattari. (1988), '1837: Of the Refrain', *A Thousand Plateaus: Capitalism and Schizophrenia*: 310–50, London: Athlone Press.
Deleuze, G. and F. Guattari. (1988), *A Thousand Plateaus: Capitalism and Schizophrenia*, trans. B. Massumi, London: Athlone Press.
Delfanti, A. (2013), *Biohackers: The Politics of Open Science*, London: PlutoPress.

References

Dennett, D. C. (1978), 'Where Am I?', in D. C. Dennett (ed.), *Brainstorms: Philosophical Essays on Mind and Psychology*: 310–23, Cambridge, MA: MIT Press.

DeNora, T. (2013), *Music Asylums: Wellbeing Through Music in Everyday Life*, Farnham, England: Ashgate.

Derrida, J. (1974), 'White Mythology: Metaphor in the Text of Philosophy', *New Literary History*, 61: 5–74.

Derrida, J. (1976), *Of Grammatology*, trans. G. C. Spivak, Baltimore, MD: Johns Hopkins University Press.

Derrida, J. (1979), *The Postcard: From Socrates to Freud and Beyond*, trans. A. Bass, Chicago, IL: Chicago University Press.

Derrida, J. (1981), 'Plato's Pharmacy', *Dissemination*: 61–171, London: Athlone.

Derrida, J. (1982), 'Différance', *Margins of Philosophy*: 3–27, Chicago: University of Chicago press.

Derrida, J. (1983), 'The Principle of Reason: The University in the Eyes of Its Pupils', *Diacritics*, 13: 3–20.

Derrida, J. (1993), *Aporias*, trans. T. Dutoit, Stanford, CA: Stanford University Press.

Derrida, J. and D. Wills. (2002), 'The Animal That Therefore I Am (More to Follow)', *Critical Inquiry*, 28 (2): 369–418.

Dewey, J. (1958), *Experience and Nature*, New York: Dover.

Dewey, J. (1980), *Art as Experience*, New York: Wideview Perigee.

Dick, P. K. (2000), *We Can Remember It for You Wholesale*, London: Millennium.

Dollens, D. (2005), 'A System of Digital-Botanic Architecture', *Leonardo*, 38 (1): 15–21.

Dollens, D. (2015), *Autopoietic-Extended Architecture: Can Buildings Think? (Phd Thesis)*, Edinburgh: The University of Edinburgh.

Dolphin Dies Being Passed around for Selfies'. (2016), *Sky News*, 18 February. Available online: http://news.sky.com/story/dolphin-dies-being-passed-around-for-selfies-10171630 (accessed 3 June 2017).

Douglas, M. (1966), *Purity and Danger: An Analysis of Concepts of Pollution and Taboo*, London: Routledge and Kegan Paul.

Dourish, P. (2001), *Where the Action Is: The Foundations of Embodied Interaction*, Cambridge, MA: MIT Press.

Dreyfus, H. L. (1991), *Being-in-the-World: A Commentary on Heidegger's Being and Time, Division I*, Cambridge, MA: MIT Press.

Drummond, H. (2016), *The Application of Cognitive Behavioural Therapy in Interactive Media*, Edinburgh, UK: Unpublished Research Report, The University of Edinburgh.

Dudley, D. (1967), *A History of Cynicism: From Diogenes to the 6th Century Ad*, Hildesheim, Germany: Georg Olms Verlagsbuchhandlung.

Dumbill, E. (2012), 'What Is Big Data? An Introduction to the Big Data Landscape'. *O'Reilly*, 11 January. Available online: https://beta.oreilly.com/ideas/what-is-big-data (accessed 11 June 2017).

Dummett, E. (2007), *Green Space and Cosmic Order: Le Corbusier's Understanding of Nature*, Edinburgh: PhD Thesis, The University of Edinburgh.

Dunn, J. E. (2012), 'Smartphones Stress Users with "Phantom" Text Messages'. *Techworld*, 12 January. Available online: http://www.techworld.com/news/personal-tech/smartphones-stress-users-with-phantom-text-messages-3329685/ (accessed 3 June 2017).

Durkheim, E. and M. Maus. (1963), *Primitive Classification*, Chicago, IL: The University of Chicago Press.

Easley, D. and J. Kleinberg. (2010), *Networks, Crowds and Markets: Reasoning About a Highly Connected World*, Cambridge, UK: Cambridge University Press.

Eco, U. (1980), *The Name of the Rose*, trans. W. Weaver, London: Vintage.

Eco, U. (1984), *Semiotics and the Philosophy of Language*, London: Macmillan.
Eliade, M. (1961), *Images and Symbols: Studies in Religious Symbolism*, trans. P. Mairet, London: Harvill Press.
Emerson, R. W. (2016), *Essays by Ralph Waldo Emerson*, New York: Dossier Press.
Factor, R. L. (1996), 'Peirce's Definition of Metaphor and Its Consequences', in V. M. Colapietro and T. M. Olshewsky (eds), *Peirce's Doctine of Signs: Theory, Application, and Connections*: 229–35, Berlin: Mouton de Gruyter.
Fairfield, J. (2009), 'The Magic Circle', *Vanderbilt Journal of Entertainment and Technology Law*, Washington & Lee Legal Studies Paper No. 2008-45.
Flanagan, M. (2009), *Critical Play: Radical Game Design*, Cambridge, MA: MIT Press.
Flatley, J. (2008), *Affective Mapping: Melancholia and the Politics of Modernism*, Cambridge, MA: Harvard University Press.
Flynn, E. (2016), '(Experimenting with) Living Architecture: A Practice Perspective', *Arq (Architectural Research Quarterley)*, 20 (1): 21–8.
Forster, E. M. (2011), *The Machine Stops*, London: Penguin.
Foucault, M. (1970), *The Order of Things: An Archaeology of the Human Sciences*, New York: Random House.
Foucault, M. (1977), *Discipline and Punish: The Birth of the Prison*, London: Penguin.
Frampton, K. (1985), *Modern Architecture: A Critical History*, London: Thames and Hudson.
Francé, R. H. (1920), *Die Pflanze Als Erfinder (the Plant as an Inventor)*, Kosmos: Stuttgart.
Freud, S. (1917), 'Mourning and Melancholia', *The Standard Edition of the Complete Psychological Works of Sigmund Freud, Volume Xiv (1914-1916): On the History of the Psycho-Analytic Movement, Papers on Metapsychology and Other Works*: 237–58, London: Hogarth Press.
Freud, S. (1991), 'Infantile Sexuality', in A. Richards (ed.), *The Penguin Freud Library, Volume 7: On Sexuality*: 88–126, Harmondsworth, Middlesex: Penguin.
Furst, L. R. (1969), *Romanticism in Perspective: A Comparative Study of Aspects of the Romantic Movements in England, France and Germany*, London: MacMillan.
Gadamer, H.-G. (1975), *Truth and Method*, trans. J. Weinsheimer, New York: Seabury Press.
Gadamer, H.-G. (2004), *Truth and Method*, trans. J. Weinsheimer and D. G. Marshall, New York: Continuum.
Gallagher, S. (1992), *Hermeneutics and Education*, Albany, NY: SUNY Press.
Gaskin, J. e. (1995), *The Epicurean Philosophers*, London: Everyman.
Gaver, B., T. Dunne and E. Pacenti. (1999), 'Design: Cultural Probes', *Interactions Magazine*, 6 (1): 21–9.
Gefter, A. (2016), 'The Evolutionay Argument against Reality'. *Quanta Magazine*, 21 April. Available online: https://www.quantamagazine.org/20160421-the-evolutionary-argument-against-reality/ (accessed 11 June 2017).
Gere, C. (2004), 'Brains-in-Vats, Giant Brains and World Brains: The Brain as Metaphor in Digital Culture', *Studies in History and Philosophy of Biological and Biomedical Sciences*, 35: 351–66.
Gibbs, S. (2015), 'Apple Co-Founder Steve Wozniak Says Humans Will Be Robots' Pets'. *Guardian*, 25 June. Available online: https://www.theguardian.com/technology/2015/jun/25/apple-co-founder-steve-wozniak-says-humans-will-be-robots-pets (accessed 3 June 2017).
Gillam, M., et al. (2009), 'The Healthcare Singularity and the Age of Semantic Medicine', in T. Hey, S. Tansley and K. Tolle (eds), *The Fourth Paradigm: Data-Intensive Scientific Discovery*: 57–64, Redmond, WA: Microsoft Research.
Gillette, J. (2005), 'Can Gardens Mean?', *Landscape Journal*, 24 (1): 85–97.
Gladwell, V. F., et al. (2013), 'The Great Outdoors: How a Green Exercise Environment Can Benefit All', *Extreme Physiology & Medicine*, 2 (3): 1–7.
Gleick, J. (1988), *Chaos: Making a New Science*, London: Heinemann.

Godwin, M. (2015), 'Ai Weiwei and Anish Kapoor Lead London Walk of Compassion for Refugees'. *Guardian*, 17 September. Available online: https://www.theguardian.com/uk-news/2015/sep/17/ai-weiwei-anish-kapoor-london-walk-refugees (accessed 3 June 2017).

Goffman, E. (1962), *Asylums: On the Social Situation of Mental Health Patients and Other Inmates*, New York: Anchor Books.

Golumbia, D. (2015), 'Bitcoin as Politics: Distributed Right-Wing Extremism', in G. Lovink, N. Tkacz and P. de Vries (eds), *Moneylab Reader: An Intervention in Digital Economy*: 117–31, Amsterdam: Institute of Network Cultures.

Goodstein, E. S. (2005), *Experience without Qualities: Boredom and Modernity*, Stanford, CA: Stanford University Press.

Gooley, T. (2014), *The Walker's Guide to Outdoor Clues and Signs*, London: Sceptre, Hodder and Stoughton.

Gould, H. (2016), 'If Pokémon Go Feels Like a Religion, That's Because It Kind of Is'. *Guardian*, 3 June 2017. Available online: https://www.theguardian.com/technology/2016/jul/12/pokemon-go-addictive-game-shares-much-with-religious-devotion (accessed 12 July 2016).

Graham, S. and S. Marvin. (2001), *Splintering Urbanism: Networked Infrastructures, Technological Mobilities and the Urban Condition*, London: Routledge.

Gray, J. (1992), *Men Are from Mars, Women Are from Venus*, London: Harper Collins.

Greenfield, A. and K. Nurri. (2013), *Against the Smart City (the City Is Here for You to Use Book 1)*, New York, NY: Do Projects.

Greenfield, A. and M. Shepard. (2007), *Situated Technologies Pamphlets 1: Urban Computing and Its Discontents*, New York: The Architectural League of New York.

Griffin, D. R. (1988), 'Introduction: The Reenchantment of Science', in D. R. Griffin (ed.), *The Reenchantment of Science: Postmodern Proposals*: 1–46, Albany, NY: SUNY Press.

Guignon, C. (1999), *The Good Life*, Indianapolis, IL: Hackett.

Hale, J. (2016), *Merleau-Ponty for Architects*, London: Routledge.

Halsted, E. (2006), 'A Shoe Is Rarely Just a Shoe: Women's Accessories and Their Psyches', in J. Petrucelli (ed.), *Longing: Psychoanalytic Musings on Desire*: 101–11, London: Karnac.

Hammer, E. (2004), 'Being Bored: Heidegger on Patience and Melancholy', *British Journal for the History of Philosophy*, 12 (2): 277–95.

Haraway, D. J. (1991), 'A Cyborg Manifesto: Science, Technology, and Socialist Feminism in the Late Twentieth Century', *Simians, Cyborgs, and Women: The Reinvention of Nature*: 149–81, London: FAb.

Haraway, D. J. (1991), *Simians, Cyborgs, and Women: The Reinvention of Nature*, London: FAb.

Hardman, M. (2014), 'Look out Behind the Bus Stop, Here Come Guerrilla Gardeners Digging up an Urban Revolution'. *The Conversation (blog site)*, 16 June. Available online: https://theconversation.com/look-out-behind-the-bus-stop-here-come-guerrilla-gardeners-digging-up-an-urban-revolution-29225 (accessed 3 June 2017).

Harries, T., et al. (2016), 'Effectiveness of a Smartphone App in Increasing Physical Activity Amongst Male Adults: A Randomised Controlled Trial', *BMC Public Health*, 16 (925): 1–10.

Harries, T., et al. (2013), 'Walking in the Wild: Using an Always-on Smartphone Application to Increase Physical Activity', in P. Kotzé, G. Marsden, G. Lindgaard, J. Wesson and M. Winckler (eds), *Lecture Notes in Computer Science, Vol 8120*: 19–36, Berlin: Springer.

Hartig, T., et al. (2003), 'Tracking Restoration in Natural and Urban Field Settings', *Journal of Environmental Psychology*, 23: 109–23.

Hauptmann, D. (2011), 'Introduction: Architecture and Mind in the Age of Communication and Information', in D. Hauptmann and W. Neidich (eds), *Cognitive Architecture: From Biopolitics to Noopolotics. Architecture and Mind in the Age of Communication and Information*: 10–45, Rotterdam: 010 Publishers.

Hawkes, T. (2003), *Structuralism and Semiotics*, London: Routledge.

Hawking, S. (1988), *A Brief History of Time: From the Big Bang to Black Holes*, New York: Bantam.

Hayles, K. N. (1996), 'Narratives of Artificial Life', in G. Robertson, M. Mash, L. Tickner, J. Bird, B. Curtis and T. Putman (eds), *Future Natural: Nature, Science, Culture*: 146–64, London: Routledge.

Hayles, N. K. (1999), *How We Became Posthuman: Virtual Bodies in Cybernetics, Literature, and Informatics*, Chicago, IL.: University of Chicago Press.

Hazarika, A. (2016), 'Labour's Comedy Therapy, the Brexit Satire Boom and the Return of the Pink Bus', *New Statesman*, 26 August–1 September: 19-.

Head, L. (2016), *Hope and Grief in the Anthropocene: Re-Conceptualising Human–Nature Relations*, Abingdon, Oxon: Routledge.

Hefferman, V. (2016), *Magic and Loss: The Internet as Art*, London: Simon and Schuster.

Heft, H. and J. L. Nasar. (2000), 'Evaluating Environmental Scenes Using Dynamic Versus Static Displays', *Environment and Behaviour*, 32 (3): 301–22.

Heidegger, M. (1962), *Being and Time*, trans. J. Macquarrie and E. Robinson, London: SCM Press.

Heidegger, M. (1971), 'Building, Dwelling, Thinking', *Poetry, Language, Thought*: 143–61, New York: Harper and Rowe.

Heidegger, M. (1974), 'The Principle of Ground', *Man and World*, 7: 207–22.

Heidegger, M. (1977), *The Question Concerning Technology and Other Essays*, trans. W. Lovitt, New York: Harper and Row.

Heidegger, M. (1995), *The Fundamental Concepts of Metaphysics: World, Finitude, Solitude*, trans. W. McNeill and N. Walker, Bloomington, Indiana: Indiana University Press.

Hendrix, J. S. and L. E. Holm. (2016), 'Aldo Rossi and the Field of the Other', in J. S. Hendrix and L. E. Holm (eds), *Architecture and the Unconscious*: 99–117, Farnham: Ashgate/Routledge.

Hesiod. (1988), *Theogony and Works and Days*, trans. M. L. West, Oxford: Oxford University Press.

Hessenfelder, S. (2013), 'The Simulation Hypothesis and Other Things I Don't Believe'. *BackReaction philosophy blog*, 28 February. Available online: http://backreaction.blogspot.co.uk/2013/02/the-simulation-hypothesis-and-other.html (accessed 11 June 2017).

Hey, T., S. Tansley and K. Tolle (eds). (2009), *The Fourth Paradigm: Data-Intensive Scientific Discovery*, Redmond, WA: Microsoft Research.

Hillier, B. (2011), 'The City as Socio-Technical System: A Spatial Reformulation in the Light of the Levels Problem and the Parallel Problem', in S. Müller Arisona, A. Gideon, J. Halatsch and P. Wonka (eds), *Digital Urban Modelling and Simulation*: 24–48, Heidelberg, Germany: Springer.

Hiltz, S. R. and M. Turoff. (1993), *The Network Nation: Human Communication Via Computer*, Cambridge, MA: MIT Press.

Hobbes, T. (1991), *Leviathan*, Cambridge, England: Cambridge University Press.

Hochschild, A. R. (2012), *The Managed Heart: Commercialization of Human Feeling*, Berkeley, CA: University of California Press.

Hodges, A. (1985), *Alan Turing: The Enigma of Intelligence*, London: Unwin Paperbacks.

Holling, C. S. (1973), 'Resiliency and Stability of Ecological Systems', *Annual Review of Ecology and Systematics*, 4: 1–23.

Houser, N. (1998), 'Introduction', in N. Houser (ed), *The Essential Peirce, Selected Philosophical Writings Volume 2 (1893-1913)*: xvii–xxxviii, Bloomington, IN: Indiana University Press.

Hughes, T. P. (2012), 'The Evolution of Large Technological Systems', in W. E. Bijker, T. P. Hughes, T. Pinch and D. G. Douglas (eds), *The Social Construction of Technological Systems: New Directions in the Sociology and History of Technology*: 45–76, Cambridge, MA: MIT Press.

Huizinga, J. (1955), *Homo Ludens: A Study of the Play Element in Culture*, Boston: Beacon Press.

Hulswit, M. (2001), 'Peirce on Causality and Causation'. *The Commens Encyclopedia: The Digital Encyclopedia of Peirce Studies. New Edition*, Pub. 120809-1715a. Available online: http://www.commens.org/encyclopedia/article/hulswit-menno-peirce-causality-and-causation (accessed 11 June 2017).

Husserl, E. (1962), *Ideas: General Introduction to Pure Phenomenology*, trans. W. R. Boyce Gibson, New York: Collier Books.

Hutchins, E. (1996), *Cognition in the Wild*, Cambridge, MA: MIT Press.

Hyde, L. (1998), *Trickster Makes This World: Mischief, Myth and Art*, New York: North Point Press.

Ihde, D. (1995), *Postphenomenology: Essays in the Postmodern Context*, Evanston, IL: Northwestern University Press.

Indick, W. (2015), *The Digital God: How Technology Will Reshape Spirituality*, Jefferson, NC: McFarland.

Isaacson, W. (2011), *Steve Jobs: The Exclusive Biography*, London: Hachette Digital.

Ison, R. (2016), 'Governing in the Anthropocene: What Future Systems Thinking in Practice?', *Systems Research and Behavioural Science*, 33 (5): 595–613.

Jabi, W. (2013), *Parametric Design for Architecture*, London: Laurence King.

Jakobson, R. and M. Halle. (1956), *Fundamentals of Language*, The Hague, Holland: Mouton.

James, J. (1989), *The Template-Makers of the Paris Basin*, Leura, Australia: West Grinstead.

James, W. (1950), *The Principles of Psychology*, New York: Dover.

James, W. (1981), *The Principles of Psychology Volume I*, Cambridge, MA: Harvard University Press.

Jameson, F. (1972), *The Prison-House of Language: A Critical Account of Structuralism and Russian Formalism*, Princeton, NJ: Princeton University Press.

Jencks, C. (1973), *Modern Movements in Architecture*, New York: Penguin.

Jencks, C. (1995), *The Architecture of the Jumping Universe, A Polemic: How Complexity Science Is Changing Architecture and Culture*, London: Academy Editions.

Jencks, C. (2005), *Iconic Building*, New York: Rizzoli.

Jencks, C. and G. Baird (eds). (1969), *Meaning in Architecture*, London: Barrie & Rockliff.

Johansen, J. M. (2011), 'Nanoarchitecture'. A Discourse Part I. Available online: http://johnmjohansen.com/Nanoarchitecture.html (accessed 11 June 2017).

Johnson, M. (1987), *The Body in the Mind: The Bodily Basis of Meaning, Imagination, and Reason*, Chicago, IL: University of Chicago Press.

Johnson, P.-A. (1994), *The Theory of Architecture: Concepts, Themes and Practices*, New York: John Wiley.

Jones, C. A. (2006), *Sensorium: Embodied Experience, Technology, and Contemporary Art*, Cambridge, MA: MIT Press.

Jones, J. C. (1970), *Design Methods: Seeds of Human Futures*, London: Wiley.

Jung, C. G. (1986), *Four Archetypes: Mother, Rebirth, Spirit, Trickster*, London: Ark.

Juslin, P. N. and D. Västfjäll. (2008), 'Emotional Responses to Music: The Need to Consider Underlying Mechanisms', *Behavioural and Brain Sciences*, 31: 559–621.

Juul, J. (2008), 'The Magic Circle and the Puzzle Piece', in S. Günzel, M. Liebe and D. Mersch (eds), *Conference Proceedings of the Philosophy of Computer Games 2008*: 56–67, Potsdam: Potsdam University Press.

Kahn, P. H. (2011), *Technological Nature: Adaptation and the Future of Human Life*, Cambridge, MA: MIT Press.

Kalay, Y. E. (2004), *Architecture's New Media: Principles, Theories, and Methods of Computer-Aided Design*, London: MIT Press.

Kaminer, T. (2017), *The Efficacy of Architecture: Political Contestation and Agency*, London: Routledge.

Kant, I. (1960), *Observations on the Feeling of the Beautiful and the Sublime*, trans. J. T. Goldthwait, Berkeley, CA: University of California Press.

Kaplan, R. and S. Kaplan. (1989), *The Experience of Nature: A Psychological Perspective*, Cambridge, UK: Cambridge University Press.

Kaplan, S. (1995), 'The Restorative Benefits of Nature: Toward an Integrative Framework', *Journal of Environmental Psychology*, 15: 169–82.

Kay, L. E. (2000), *Who Wrote the Book of Life? A History of the Genetic Code*, Stanford: Stanford University Press.

Kayser, M., et al. (2013), 'Silk Pavillion'. *Mediated Matter*. Available online: http://matter.media.mit.edu/environments/details/silk-pavillion (accessed4 June 2017).

Keeton, W. T. (1972), *Biological Science*, New York: Norton.

Kellert, S. R. (2008), 'Dimensions, Elements, and Attributes of Biophilic Design', in S. R. Kellert, J. Heerwagen and M. Mador (eds), *Biophilic Design: The Theory, Science and Practice of Bringing Buildings to Life*: 3–19, Hoboken, NJ: Wiley.

Kellert, S. R., J. Heerwagen and M. Mador. (2008), *Biophilic Design: The Theory, Science and Practice of Bringing Buildings to Life*, Hoboken, NJ: Wiley.

Kermode, M. (2016), 'Zootropolis Review: Disney's Animated Odd Couple Has a Perfect Chemistry'. *Guardian*, 27 March. Available online: https://www.theguardian.com/film/2016/mar/27/zootropolis-review-disney-animated-odd-couple-mark-kermode (accessed 3 June 2017).

Kidder, J. L. (2012), 'Parkour, the Affective Appropriation of Urban Spaces, and the Real/Virtual Dialectic', *City and Community*, 11 (3): 229–53.

Kitchin, R. (2014), 'The Real-Time City? Big Data and Smart Urbanism', *GeoJournal*, 79: 1–14.

Klanten, R., J. Bowman and S. Ehmann. (2014), *The Outsiders: New Outdoor Creativity*, Berlin: Gestalten.

Klooster, T. (2009), *Smart Surfaces and Their Application in Architecture and Design*, Basel, Switzerland: Birkhäuser.

Kohn, E. (2013), *How Forests Think: Anthropology Beyond the Human*, Berkeley, CA: University of California Press.

Kolakowski, M. (2016), 'Modernism or Tradition in Low-Technology? A Humanistic Perspective on the Architecture of Paulina Wojciechowska', *ACEE Architecture Civil Engineering* (1): 21–34.

Kübler-Ross, E. (2014), *On Death and Dying: What the Dying Have to Teach Doctors, Nurses, Clergy and Their Own Families*, New York: Scribner.

Lakoff, G. (2003), *Women, Fire, and Dangerous Things: What Categories Reveal About the Mind*, Chicago: University of Chicago Press.

Lakoff, G. and M. Johnson. (1980), *Metaphors We Live By*, Chicago, IL: University of Chicago Press.

Langton, C. G., ed. (1997), *Artifical Life: An Overview*, Cambridge, MA: MIT Press.

Langton, C. (1989), 'Artificial Life'. Available online: http://90.146.8.18/de/archiv_files/19931/1993_025.pdf (accessed 3 June 2017).

Lefebvre, H. (1991), *The Production of Space*, trans. D. Nicholson-Smith, Oxford, UK: Blackwell.

Lefebvre, H. (2004), *Rhythmanalysis: Space, Time and Everyday Life*, trans. S. Elden and G. Moore, London: Continuum.

Leonard, P. (2013), *Literature after Globalization: Textuality, Technology and the Nation-State*, London: Bloomsbury.

Lévi-Strauss, C. (1963), *Structural Anthropology 1*, trans. C. Jacobson and B. G. Schoepf, New York: Basic Books.
Levy, S. (1992), *Artificial Life: The Quest for a New Creation*, London, England: Penguin.
Lewis, C. S. (1950), *The Lion, the Witch and the Wardrobe*, London: Geoffrey Bles.
Lewis, S. L. and M. A. Maslin. (2015), 'Defining the Anthropocene', *Nature*, 519: 170–80.
Louv, R. (2005), *Last Child in the Woods: Saving Our Children from Nature-Deficit Disorder*, London: Atlantic Books.
'Lovers Abramović & Ulay Walk the Length of the Great Wall of China from Opposite Ends, Meet in the Middle and Breakup'. (2015), *Kickass Trips*, 14 January. Available online: http://kickasstrips.com/2015/01/lovers-abramovic-ulay-walk-the-length-of-the-great-wall-of-china-from-opposite-ends-meet-in-the-middle-and-breakup/ (accessed 3 June 2017).
Lucas, G. (2017), 'Yoda Quotes'. *Genius Album Guides*. Available online: https://genius.com/George-lucas-yoda-quotes-annotated (accessed4 June 2017).
Luhmann, N. (1995), *Social Systems*, trans. J. Bednarz and D. Baecker, Stanford, CA: Stanford University Press.
Lyn, G., ed. (2004), *Folding in Architecture (Revised Edition)*, Chichester, England: Wiley-Academy.
Lynn, G. (1999), *Animate Form*, New York: Princeton Architectural Press.
Lynn, G. (2002), *Architecture for an Embryologic Housing*, Berlin: Birkhauser Verlag AG.
Macdonald, A. (2001), *Structure in Architecture*, Abingdon, Oxon: Taylor and Francis.
Macdonald, P. (2016), 'Change in Glen Feshie: Environmental Change in a Dynamic Cairngorms Landscape', *The Nature of Scotland* (24): 12–7.
Macdonald, P. and A. Macdonald. (2017), 'Rephotography in the Scottish Highlands: Cairngorms & Morvern'. *Aerographica*, 7 June. Available online: http://aerographica.org/re-photography-in-the-cairngorms-and-morvern-scottish-highlands/ (accessed 11 June 2017).
Macnaghten, P. and J. Urry. (1998), *Contested Natures*, London: Sage.
Malamud, R. (2007), 'Introduction: Famous Animals in Modern Culture', in R. Malamud (ed.), *A Cultural History of Animals in the Modern Age*: 1–26, Oxford: Berg.
Mallgrave, H. F. and D. Goodman. (2011), *An Introduction to Architectural Theory: 1968 to the Present*, Chichester, England: Wiley-Blackwell.
March, L., ed. (1976), *The Architecture of Form*, London: Cambridge University Press.
Marcuse, H. (1987), *Eros and Civilisation: A Philosophical Inquiry into Freud*, London: Routledge and Kegan Paul.
Markoš, A. (2002), *Readers of the Book of Life: Contextualizing Developmental Evolutionary Biology*, New York, NY: Oxford University Press.
Markowitz, E. M. and A. F. Shariff. (2012), 'Climate Change and Moral Judgement', *Nature Climate Change*, 2: 243–7.
Marx, K. (1977), 'Capital', in D. McLellan (ed.), *Karl Marx: Selected Writings*: 415–507, Oxford: Oxford University Press.
Maturana, H. (1988), 'Reality: The Search for Objectivity or the Quest for a Compelling Argument', *The Irish Journal of Psychology*, 9 (1): 25–82.
Maturana, H. and F. G. Varela. (1980), *Autopoiesis and Cognition*, Dordrecht: Reidel.
Mavros, P., R. Coyne, J. Roe and P. Aspinall. (2012), 'Engaging the Brain: Implications of Mobile Eeg for Spatial Representation', in H. Achten, J. Pavlicek, J. Hulin and D. Matejdan (eds), *Digital Physicality Proceedings of the 30th Ecaade Conference*: 657–65, Czech Technical University in Prague: Molab.
Mayer-Schonberger, V. and K. Cukier. (2013), *Big Data: A Revolution That Will Transform How We Live, Work, and Think*, Boston, MA: Eamon Dolan.
McCarthy, J. and P. Wright. (2004), *Technology as Experience*, Cambridge, MA: MIT Press.

McCullough, M. (2004), *Digital Ground: Architecture, Pervasive Computing, and Environmental Knowing*, Cambridge, MA: MIT Press.

Mccullough, M. (2013), *Ambient Commons: Attention in the Age of Embodied Information*, Cambridge, MA: MIT Press.

McKibben, B. (1989), *The End of Nature*, New York, NY: Random House.

McKibben, B. (2002), 'The End of Nature', in R. Finch and J. Elder (eds), *The Norton Book of Nature Writing: College Edition*: 1120–30, New York: W. W. Norton.

McLuhan, M. (1994), *Understanding Media, the Extensions of Man*, Cambridge, MA: MIT Press.

Mendenhall, A. (2015), 'The Classical Liberalism of Ralph Waldo Emerson'. *The Literary Lawyer: A Forum for the Legal and Literary Communities*, 7 January. Available online: https://allenmendenhallblog.com/2015/01/07/the-classical-liberalism-of-ralph-waldo-emerson/ (accessed 11 June 2017).

Mezes, A. (2012), 'Your Smartphone May Be Stressing You Out'. *USA Today*, 13 January. Available online: http://usatoday30.usatoday.com/news/health/story/health/story/2012-01-13/Your-smartphone-may-be-stressing-you-out/52529514/1 (accessed 11 June 2017).

Miedaner, T. (1981), 'The Soul of the Mark III Beast (and Reflection)', in D. R. Hofstadter and D. C. Dennett (eds), *The Mind's Eye: Fantasies and Reflections on Self and Soul*: 109–15, New York: Basic Books.

Miller, F. and L. Colloca. (2010), 'Semiotics and the Placebo Effect', *Perspectives in Biology and Medicine*, 53 (4): 509–16.

Miller, H. J. and M. F. Goodchild. (2015), 'Data-Driven Geography', *GeoJournal*, 80: 449–61.

Mitchell, W. J. (1977), *Computer-Aided Architectural Design*, New York, NY: John Wiley and Sons.

Mitchell, W. J. (2003), *Me++: The Cyborg Self and the Networked City*, Cambridge. MA: MIT Press.

Mitchell, W. J. T. (1995), *Picture Theory: Essays on Verbal and Visual Representation*, Chicago: University of Chicago Press.

Mitchell, W. J. T. (2002), 'Imperial Landscape', in W. J. T. Mitchell (ed.), *Landscape and Power*: 5–34, Chicago: University of Chicago Press.

Mitchell, W. J. T. (2002), *Landscape and Power*, Chicago: University of Chicago Press.

Moholy-Nagy, S. (1950), *Moholy-Nagy: Experiment in Totality*, Cambridge, MA: MIT Press.

Moravec, H. P. (1988), *Mind Children: The Future of Robot and Human Intelligence*, Cambridge, MA: Harvard University Press.

Moretti, F. (2013), *Distant Reading*, London: Verso.

Morozov, E. (2011), *The Net Delusion: How Not to Liberate the World*, London: Allen Lane.

Morrissey, T. J. and R. Wunderlich. (1983), 'Death and Rebirth in Pinocchio', *Children's Literature*, 11: 64–75.

Morrow, B. (2016), 'Does Donald Trump Run His Own Twitter Account? 5 Fast Facts You Need to Know'. *Heavy*, 2 June. Available online: http://heavy.com/news/2016/06/does-donald-trump-run-his-twitter-account-page-operate-use-post-write-update-campaign-controversy-retweet-tweet-white-supremacist-bots-followers-first/ (accessed 11 June 2017).

Navlakha, S. and Z. Bar-Joseph. (2015), 'Distributed Information Processing in Biological Systems', *Communication of the ACM*, 58 (1): 94–102.

Negroponte, N. (1995), *Being Digital*, London: Hodder and Stoughton.

Negroponte, N. (1998), 'Beyond Digital', *Wired Magazine*, 6 (12).

Newell, A. and H. Simon. (1972), *Human Problem Solving*, Englewood Cliffs, NJ: Prentice-Hall.

Ngai, S. (2012), *Our Aesthetic Categories: Zany, Cute, Interesting*, Cambridge, MA: Harvard University Press.

Nicholson, C. J. (2004), 'Elegance and Grass Roots: The Neglected Philosophy of Frederick Law Olmsted', *Transactions of the Charles S. Peirce Society*, 40 (2): 335–48.

Nickerson, R. S. (1998), 'Confirmation Bias: A Ubiquitous Phenomenon in Many Guises', *Review of General Psychology*, 2 (2): 175–220.

Nietzsche, F. W. (1977), *The Nietzsche Reader*, trans. R. J. Hollingdale, London: Penguin.

Noddings, N. (1984), *Caring: A Feminine Approach to Ethics and Moral Education*, Berkeley, CA: University of California Press.

Norberg-Schulz, C. (1980), *Genius Loci: Towards a Phenomenology of Architecture*, New York: Rizzoli.

Norris, C. (1987), *Derrida*, Cambridge, MA: Harvard University Press.

North, G. (2014), 'Thoreau's Walden: Phony Testament of the Greens'. *Gary North's Specific Answers*, 18 April. Available online: https://www.garynorth.com/public/12347.cfm (accessed 4 June 2017).

Novak, M. (1994), 'Liquid Architectures in Cyberspace', in M. Benedikt (ed.), *Cyberspace: First Steps*: 225–54, Cambridge, MA: MIT Press.

O'Leary, S. (2004), 'Cyberspace as Sacred Space: Communicating Religion on Computer Networks', in L. L. Dawson and D. E. Cowan (eds), *Religion Online: Finding Faith on the Internet*: 37–58.

Ofcom. (2016), *The Communications Market Report*, London: Ofcom.

Ogden, C. K. and I. A. Richards. (1989), *The Meaning of Meaning: A Study of the Influence of Language Upon Thought and of the Science of Symbolism*, Orlando, FL: Harvest.

Packard, V. (2007), *The Hidden Persuaders*, Brooklyn, NY: Ig Publishing.

Palmesino, A., J. Rönnskog and E. Turpin. (2013), 'Matters of Observation on Architecture in the Anthropocene', in E. Turpin (ed.), *Architecture in the Anthropocene: Encounters among Design, Deep Time, Science and Philosophy*: 15–24, Ann Arbor, MI: Open Humanities Press.

Papadakis, A., C. Cooke and A. Benjamin (eds). (1989), *Deconstruction: Omnibus Volume*, London: Academy Editions.

Parikka, J. (2015), *A Geology of Media*, Minneapolis, MN: University of Minnesota Press.

Parker, F. (2017), 'Brexit Terms Should Be 'Reset', Labour's Shadow Brexit Secretary Says'. *Metro*, 9 June. Available online: http://metro.co.uk/2017/06/09/brexit-terms-should-be-reset-labours-shadow-brexit-secretary-says-6696996/ (accessed 10 June 2017).

Pavement Lights Guide 'Smartphone Zombies''. (2017), *BBC News*, 16 February. Available online: http://www.bbc.co.uk/news/technology-38992653 (accessed 3 June 2017).

Peirce, C. S. (1932), *The Collected Papers of Charles Sanders Peirce. Vol. 2 Elements of Logic*, Cambridge: Harvard University Press.

Peirce, C. S. (1935), *The Collected Papers of Charles Sanders Peirce. Vol. 5 Pragmatism and Pragmaticism*, Cambridge: Harvard University Press.

Peirce, C. S. (1992), *The Essential Peirce, Selected Philosophical Writings Volume 1 (1867-1893)*, Bloomington, IN: Indiana University Press.

Peirce, C. S. (1998), *The Essential Peirce, Selected Philosophical Writings Volume 2 (1893-1913)*, Bloomington, IN: Indiana University Press.

Peirce, C. S. (1998), 'Immortality in the Light of Synechism', in N. Houser (ed.), *The Essential Peirce, Selected Philosophical Writings Volume 2 (1893-1913)*: 1–3, Bloomington, IN: Indiana University Press.

Peirce, C. S. (1998), 'The Maxim of Pragmatism', in N. Houser (ed), *The Essential Peirce, Selected Philosophical Writings Volume 2 (1893-1913)*: 133–44, Bloomington, IN: Indiana University Press.

Peirce, C. S. (1998), 'The Neglected Argument for the Reality of God', in N. Houser (ed.), *The Essential Peirce, Selected Philosophical Writings Volume 2 (1893-1913)*: 434–50, Bloomington, IN: Indiana University Press.

Peirce, C. S. (1998), 'What Is a Sign?', in N. Houser (ed.), *The Essential Peirce, Selected Philosophical Writings Volume 2 (1893-1913)*: 4–10, Bloomington, IN: Indiana University Press.

Peirce, C. S. (1998), 'What Makes a Reasoning Sound?', in N. Houser (ed.), *The Essential Peirce, Selected Philosophical Writings Volume* 2 (1893-1913): 242–57, Bloomington, IN: Indiana University Press.

Pels, P. (2003), 'Introduction', in B. Meyer and P. Pels (eds), *Magic and Modernity: Interfaces of Revelation and Concealment*: 1–38, Stanford, CA: Stanford University Press.

Penrose, R. (1989), *The Emperor's New Mind: Concerning Computers, Minds, and the Laws of Physics*, London: Vintage.

Perez-Gomez, A. (2016), *Attunement: Architectural Meaning after the Crisis of Modern Science*, Cambridge, MA: MIT Press.

Pettigrew, D. E. (1996), 'Peirce and Derrida: From Sign to Sign', in V. M. Colapietro and T. M. Olshewsky (eds), *Peirce's Doctrine of Signs: Theory, Applications, and Connections*: 365–78, Berlin: Mouton de Gruyter.

Pevsner, N. (2005), *Pioneers of Modern Design: From William Morris to Walter Gropius*, New Haven, CN: Yale University Press.

Piaget, J. (1971), *Biology and Knowledge*, Edinburgh, UK: Edinburgh University Press.

Picon, A. (2015), *Smart Cities: A Spatialised Intelligence*, London: Wiley.

Plant, S. (1998), *Zeros and Ones: Digital Women and the New Technoculture*, London: Fourth Estate.

Plato. (1995), *Statesman*, trans. J. Annas and R. Waterfield, Cambridge, England: Cambridge University Press.

Pliny, T. E. (1991), *Natural History: A Selection*, trans. J. F. Healy, London: Penguin.

Plotinus. (1948), *The Essence of Plotinus: Extracts from the Six Enneads and Porphyry's Life of Plotinus*, trans. S. Mackenna, New York: Oxford University Press.

Pooley, C., et al. (2011), *Understanding Walking and Cycling: Summary of Key Findings and Recommendations*, Lancaster, England: Lancaster University.

Preziosi, D. (1979), *Architecture, Language and Meaning: Origins of the Built World and Its Semiotic Organization*, The Hague, The Netherlands: Mouton.

Prüfer, T. and S. Mondial. (2015), '"Ingress": When Google Plays Games in a Concentration Camp'. *Zeit Magazin*, 1 July. Available online: http://www.zeit.de/zeit-magazin/leben/2015-07/ingress-smartphone-game-google-niantic-labs-nazis-concentration-camp (accessed 3 June 2017).

Pullan, W. (2015), 'Agon in Urban Conflict: Some Possibilities', in H. Steiner and M. Sternberg (eds), *Phenomenologies of the City: Studies in the History and Philosophy of Architecture*: 213–24, Farnham, England: Ashgate.

Radford, A. D. and G. Stevens. (1986), *Computer Aided Design Made Easy: A Comprehensive Guide for Architects and Designers*, Mcgraw-Hill.

Ransom, B. (2016), 'The Sixth Element: Diy Cyborgs and the Hive Minds of Social Media', in C. N. Terranova and M. Tromble (eds), *The Routledge Companion to Biology in Art and Architecture*: 458–66, London: Routledge.

Reddy, M. (1979), 'The Conduit Metaphor: A Case of Frame Conflict in Our Language About Language', in A. Ortony (ed.), *Metaphor and Thought*: 284–324, Cambridge: Cambridge University Press.

Redmond, A., A. Hore, M. Alshawi and R. West. (2012), 'Exploring How Information Exchanges can be Enhanced Through Cloud BIM', *Automation in Construction*, 24: 175–83.

Reynolds, L. A. and P. Sawyer. (1959), 'Folk Medicine and the Four Fairies of *A Midsummer Night's Dream*', *Shakespeare Quarterly*, 10 (4): 513–21.

References

Rheingold, H. (1993), *The Virtual Community: Homesteading on the Electronic Frontier*, Reading, MA: Addison-Wesley.

ricca_riot. (2015), 'Interstellar Transmissions'. *Archive of Our Own*, Chapter 2. Available online: http://archiveofourown.org/works/5496170/chapters/12742442?view_adult=true (accessed 4 June 2017).

Ricoeur, P. (1971), 'What Is a Text: Explanation and Interpretation', in D. M. Rasmussen (ed.), *Myth-Symbolic Language and Philosophical Anthropology*: 135–50, The Hague, Netherlands: Martinus Nijhoff.

Rittel, H. and M. Webber. (1973), 'Dilemmas in a General Theory of Planning', *Policy Sciences*, 4: 155–69.

Robins, K. (1995), 'Cyberspace and the World We Live In', in M. Featherstone and R. Burrows (eds), *Cyberspace, Cyberbodies, Cyberpunk: Cultures of Technological Embodiment*: 135–56, London: Sage.

Roe, J. and P. Aspinall. (2011), 'The Restorative Benefits of Walking in Urban and Rural Settings in Adults with Good and Poor Mental Health', *Health and Place*, 17: 103–13.

Roe, J. J., P. A. Aspinall, P. Mavros and R. Coyne. (2013), 'Engaging the Brain: The Impact of Natural Versus Urban Scenes Using Novel Eeg Methods in an Experimental Setting', *Environmental Sciences*, 1 (2): 93–104.

Roethlisberger, F. J. and W. J. Dickson. (1939), *Management and the Worker*, Cambridge, MA: Harvard University Press.

Rose, J. F. P. (2016), *The Well-Tempered City: What Modern Science, Ancient Civilizations, and Human Nature Teach Us About the Future of Urban Life*, London: HarperCollins.

Rosen, L. D. (2012), *Idisorder: Understanding Our Obsession with Technology and Overcoming Its Hold on Us*, New York: Palgrave MacMillan.

Rosenberg, C. (2013), 'Improving Photo Search: A Step across the Semantic Gap'. *Google Research Blog*, 12 June. Available online: https://research.googleblog.com/2013/06/improving-photo-search-step-across.html (accessed 3 June 2017).

Rowley, J. (2007), 'The Wisdom Hierarchy: Representations of the Dikw Hierarchy', *Journal of Information Science*, 33 (2): 163–80.

Ruskin, J. and C. Wilmer (ed.). (1985), *Unto This Last: And Other Writings*, London: Penguin.

Russell, B. (1945), *A History of Western Philosophy*, New York, NY: Simon and Schuster.

Russell, J. A. and G. Pratt. (1980), 'A Description of the Affective Quality Attributed to Environments', *Journal of Personality and Social Psychology*, 38 (2): 311–22.

Saussure, F. d. (1983), *Course in General Linguistics*, trans. R. Harris, London: Duckworth.

Saville, B. K., A. Gisbert, J. Kopp and C. Telesco. (2010), 'Internet Addiction and Delay Discounting in College Students', *The Psychological Record*, 60: 273–86.

Sax, B. (2007), 'The Cosmic Spider and Her Worldwide Web: Sacred and Symbolic Animals in the Era of Change', in R. Malamud (ed), *A Cultural History of Animals in the Modern Age*: 27–48, Oxford: Berg.

Sayre, F. (1938), *Diogenes of Sinope*, Baltimore: J. H. Furst Company.

Schön, D. (1979), 'Generative Metaphor: A Perspective on Problem-Setting in Social Policy', in A. Ortony (ed.), *Metaphor and Thought*: 254–83, Cambridge: Cambridge University Press.

Schulz, K. (2011), 'What Is Distant Reading?'. *The New York Times*, 24 June. Available online: http://www.nytimes.com/2011/06/26/books/review/the-mechanic-muse-what-is-distant-reading.html?_r=0 (accessed 11 June 2017).

Schumacher, P. (2008), 'Parametricism as Style: Parametricist Manifesto'. *Writings*, Presentation at the Dark Side Club, 11th Architecture Biennale, Venice. Available online: http://www.patrikschumacher.com/Texts/Parametricism as Style.htm (accessed 11 June 2013).

Schumacher, P. (2009), 'Parametricism: A New Global Style for Architecture and Urban Design', *Architectural Design*, 79 (4): 14–23.

Sebeok, T. A. (1999), *Signs: An Introduction to Semiotics*, Toronto: University of Toronto Press.

Selhub, E. M. and A. C. Logan. (2012), *Your Brain on Nature: The Science of Nature's Influence on Your Health, Happiness and Vitality*, New York: John Wiley.

Sellars, J. (1999), 'The Point of View of the Cosmos: Deleuze, Romanticism, Stoicism', *Pli (The Warwick Journal of Philosophy)*, 8: 1–24.

Serres, M. and F. McCarren. (1992), 'The Natural Contract', *Critical Inquiry*, 19 (1): 1–21.

Shannon, C. E. and W. Weaver. (1963), *The Mathematical Theory of Communication*, Urbana, IL: The University of Illinois Press.

Shapiro, E. and Y. Benenson. (2006), 'Bringing DNA Computers to Life', *Scientific American*, 294: 44–51.

Sharov, A. A. and T. Vehkavaara. (2014), 'Protosemiosis: Agency with Reduced Representation Capacity', *Biosemiotics*, 8 (1): 103–23.

Sharr, A. (2006), *Heidegger's Hut*, Cambridge, MA: MIT Press.

Shaw, J. (1996), 'The Dis-Embodied Re-Embodied Body', *Kunstforum, Die Zukunft des Körpers I*, 132.

Sheldrake, R. (2009), *Morphic Resonance: The Nature of Formative Causation*, Rochester, VT: Park Street Press.

Shelley, M. (1992), *Frankenstein*, Basingstoke: Macmillan.

Shepard, P. (1996), *The Others: How Animals Made Us Human*, Washington, DC: Island Press.

Simon, H. (1969), *The Sciences of the Artificial*, Cambridge, MA: MIT Press.

Simons, D. J. and C. F. Chabris. (1999), 'Gorillas in Our Midst: Sustained Inattentional Blindness for Dynamic Events', *Perception*, 28: 1059–74.

'Smartphones Can Increase Stress Levels, Study Says'. (2012), *Huffpost*, 12 January. Available online: http://www.huffingtonpost.com/2012/01/12/smartphones-cause-stress_n_1202924.html (accessed 3 June 2017).

Smith, A. (1984), *The Theory of Moral Sentiments*, Indianapolis: Liberty Fund.

Smyth, R. (2017), 'Happy Planet?', *BBC Wildlife*, 35 (6): 31–4.

Snodgrass, A. and R. Coyne. (2006), *Interpretation in Architecture: Design as a Way of Thinking*, London: Routledge.

Sölin, S. (2016), 'Do Glaciers Speak: The Political Aesthetics of Vo/Ice', in J. Thorpe, S. Rutherford and L. A. Sandberg (eds), *Methodological Challenges in Nature-Culture and Environmental History Research*: 13–30, New York: Routledge.

Sonesson, G. (2015), 'Phenomenology Meets Semiotics: Two Not So Very Strange Bedfellows at the End of Their Cinderella Sleep', *Metodo. International Studies in Phenomenology and Philosophy*, 3 (1): 41–62.

Sprigge, T. L. S. (1997), 'Pantheism', *The Monist*, 80 (2): 191–217.

Stallabras, J. (1996), *Gargantua: Manufactured Mass Culture*, London: Verso.

Stapleton, M. (2012), 'Feeling the Strain: Predicting the Third Dimension of Core Affect', *Behavioural and Brain Sciences*, 35 (3): 166–7.

Steadman, P. (1983), *Architectural Morphology: An Introduction to the Geometry of Building Plans*, London: Pion.

Steadman, P. (2008), *Evolution of Designs: Biological Analogy in Architecture and the Applied Arts*, London: Routledge.

Stout, C. T. and R. Kline. (2015), 'Racial Salience, Viability, and the Wilder Effect: Evaluating Polling Accuracy for Black Candidates', *Public Opinion Quarterly*, 79 (4): 994–1014.

Straus, E. W. M. (1966), *Phenomenological Psychology: The Selected Papers of Erwin W. Straus*, trans. E. Eng, London: Tavistock.
Street, C. (2010), *London's Ley Lines: Pathways of Enlightenment*, Earthstars Publishing.
Sturluson, S. (2006), *The Prose Edda: Norse Mythology*, trans. J. L. Byock, London: Penguin.
Sugiyama, T. and C. Ward Thompson. (2007), 'Older People's Health, Outdoor Activity and Supportiveness of Neighbourhood Environments', *Landscape and Urban Planning*, 83: 168–75.
Szerszynski, B. (2012), 'The End of the End of Nature: The Anthropocene and the Fate of the Human', *Oxford Literary Review*, 34 (2): 165–84.
Szerszynski, B. (2017), 'Gods of the Anthropocene: Geo-Spiritual Formations in the Earth's New Epoch', *Theory, Culture and Society*, 34 (2–3): 253–75.
Terranova, C. N. and M. Tromble (eds) (2016), *The Routledge Companion to Biology in Art and Architecture*, London: Routledge.
Thaler, A. D., et al. (2012), 'Digital Environmentalism: Tools and Strategies for the Evolving Online Ecosystem', in E. Gallagher (ed.), *Environmental Leadership: A Reference Handbook*: 364–72, Los Angeles, CA: Sage.
Thomas, S. (2013), *Technobiophilia: Nature and Cyberspace*, London: Bloomsbury.
Thompson, D. A. W. (2014), *On Growth and Form*, Cambridge, England: Cambridge University Press.
Thoreau, H. D. (1989), *Walden: Or, Life in the Woods*, Princeton, NJ: Princeton Paperbacks.
Thornton, T. F. and Y. Malhi. (2016), 'The Trickster in the Anthropocene', *The Anthropocene Review*, 3 (3): 201–4.
Tiersky, R. (2016), 'Ralph Waldo Emerson and Donald Trump: Will Power Educate the Potentate?'. *Huffington Post*, 17 May. Available online: http://www.huffingtonpost.com/ronald-tiersky/ralph-waldo-emerson-and-d_b_9998844.html (accessed 11 June 2017).
Tolkien, J. R. R. (2005), *The Lord of the Rings*, London: Harper Collins.
Topolinski, S. and R. Reber. (2010), 'Gaining Insight into the "Aha" Experience', *Current Directions in Psychological Science*, 19 (6): 402–5.
Travis, C. and P. Holm. (2016), 'The Digital Environmental Humanities – What Is It and Why Do We Need It? The NorFish Project and Smartcity Lifeworlds', in C. Travis and A. von Lünen (eds), *The Digital Arts and Humanities: Neography, Social Media and Big Data Integrations and Applications*: 187–204, AG Switzerland: Springer.
Treib, M. (2002), 'Must Landscapes Mean?', in S. Swaffield (ed.), *Theory in Landscape Architecture: A Reader*, 89–102, Philadelphia, PA: University of Pennsylvania Press.
Treib, M. e. (2011), *Meaning in Landscape Architecture and Gardens: Four Essays, Four Commentaries*, London: Abingdon, Oxon.
Turing, A. M. (1995), 'Computing Machinery and Intelligence', in E. A. Feigenbaum and J. Feldman (eds), *Computers and Thought*: 11–35, Cambridge, MA: MIT Press.
Turkle, S. (2011), *Alone Together: Why We Expect More from Technology and Less from Each Other*, New York: Basic Books.
Turner, F. (2006), *From Counterculture to Cyberculture: Stewart Brand, the Whole Earth Network, and the Rise of Digital Utopianism*, Chicago, IL: The University of Chicago Press.
Turner, F. (2013), *The Democratic Surround: Multimedia and American Liberalism from World War Ii to the Psychedelic Sixties*, Chicago, IL: The University of Chicago Press.
Ulrich, R. S., et al. (1991), 'Stress Recovery During Exposure to Natural and Urban Environments', *Journal of Environmental Psychology*, 11: 201–30.
United Nations / Framework Convention on Climate Change. (2015), *Adoption of the Paris Agreement, 21st Conference of the Parties*, Paris: United Nations.
van Mensvoort, K. and H.-J. Grievink. (2011), *Next Nature: Nature Changes Along With Us*, Barcelona, Spain: Actar.

Vitruvius, P. (1960), *Vitruvius: The Ten Books on Architecture*, trans. M. H. Morgan, New York: Dover Publications.
von Schnurbein, S. (2000), 'The Function of Loki in Snorri Sturluson's "Edda"', *History of Religions*, 40 (2): 109–24.
Walking 'Cuts Risk of Stroke in Men'. (2013), *Evening Standard*, 14 November. Available online: http://www.standard.co.uk/news/health/walking-cuts-risk-of-stroke-in-men-8940563.html (accessed 3 June 2017).
Walsh, C. (2015), 'The Hollow Mountain: Voice in the Performance of Maria Fusco's "Master Rock"'. *Map Magazine*. Available online: http://mapmagazine.co.uk/9890/hollow-mountain/ (accessed 11 June 2017).
Wang, T. (2013), 'Big Data Needs Thick Data'. *Ethnography Matters*, 13 May. Available online: http://ethnographymatters.net/blog/2013/05/13/big-data-needs-thick-data/ (accessed 11 June 2017).
Ward Thompson, C. (2010), 'Landscape Quality and Quality of Life', in C. Ward Thompson, P. Aspinall and S. Bell (eds), *Innovative Approaches for Researching Landscape and Health: Open Space: People Space 2*: 230–55, Abingdon, England: Routledge.
Ward Thompson, C., P. Aspinall and S. Bell (eds) (2010), *Innovative Approaches for Researching Landscape and Health: Open Space: People Space 2*, Abingdon, England: Routledge.
Ward Thompson, C., et al. (2012), 'More Green Space Is Linked to Less Stress in Deprived Communities: Evidence from Salivary Cortisol Patterns', *Landscape and Urban Planning*, 105: 221–9.
Watkins, A. (1922), *Early British Trackways*, London: Simkin, Marshall, Hamilton, Kent and Co.
Watsuji, T. (1996), *Watsuji Tetsuro's Rinrigaku: Ethics in Japan*, trans. Y. Seisaku and R. E. Carter, Albany: State University of New York Press.
Waugh, E. (1951), *Brideshead Revisited: The Sacred and Profane Memories of Captain Charles Ryder*, Harmondsworth, England: Penguin.
Weber, M. (1946), 'Science as a Vocation', in H. H. Gerth and C. Wright Mills (eds), *Max Weber: Essays in Sociology*: 129–56, New York: Oxford University Press.
Weiner, N. (1950), *The Human Use of Human Beings*, Cambridge, MA: Da Capo Press.
Wells, N. M. and K. S. Lekies. (2006), 'Nature and the Life Course: Pathways from Childhood Nature Experiences to Adult Environmentalism', *Children, Youth and Environments*, 16 (1): 1–24.
Whyte, I. B., ed. (2003), *Modernism and the Spirit of the City*, London: Routledge.
Wigley, M. (2001), 'Network Fever', *Grey Room*, 4: 82–122.
Willis, K. S. (2016), *Netspaces: Space and Place in a Networked World*, Farnham, UK: Ashgate.
Wills, J. (2002), 'Digital Dinosaurs and Artificial Life: Exploring the Culture of Nature in Computer and Video Games', *Cultural Values*, 6 (4): 395–417.
Wilson, E. O. (1984), *Biophilia*, Cambridge, MA: Harvard University Press.
Wilson, R. A. and A. Clark. (2009), 'How to Situate Cognition: Letting Nature Take Its Course', in M. Aydede and P. Robbins (eds), *The Cambridge Handbook of Situated Cognition*: 55–77, Cambridge, England: Cambridge University Press.
Winograd, T. and F. Flores. (1986), *Understanding Computers and Cognition: A New Foundation for Design*, Reading, MA: Addison-Wesley.
Winter, D. (2000), 'Ports of Gaol: "Eggs Files" for Dragons Feeling for Europe's Navigators – Spiritual Destiny of Port-U-Graal, Europe and the Great Dragon Line'. *ImplosionGroup*, August. Available online: http://www.bibliotecapleyades.net/sumer_anunnaki/reptiles/reptiles29.htm (accessed 11 June 2017).
Wolstenholme, S. (1997), 'Introduction', in S. Wolstenholme (ed.), *The Wonderful Wizard of Oz by Frank L. Baum*: ix–xliii, Oxford: Oxford University Press.
Woodbury, R. (2010), *Elements of Parametric Design*, London: Routledge.

References

Wundt, W. (1897), *Outlines of Psychology*, trans. C. H. Judd, Leipzig: W. Engelmann.

Wyndham, J. (1971), *The Day of the Triffids*, London: Penguin.

Zalasiewicz, J., et al. (2008), 'Are We Now Living in the Anthropocene?', *GSA (Geological Society of America) Today*, 18 (2): 4–8.

Zetter, K. (2008), 'Humans Are Just Machines for Propagating Memes (Interview with Susan Blackmore)'. *Wired Magazine*, 29 February. Available online: https://www.wired.com/2008/02/ted-blackmore/?currentPage=all (accessed 11 June 2017).

Zuckerberg, M. (2016), 'Can We Cure All Diseases in Our Children's Lifetime?'. *Facebook Notes*, 21 September. Available online: https://www.facebook.com/notes/mark-zuckerberg/can-we-cure-all-diseases-in-our-childrens-lifetime/10154087783966634 (accessed 4 June 2017).

INDEX

Index

Index